EPISODES IN NINETEENTH AND TWENTIETH CENTURY EUCLIDEAN GEOMETRY

NEW MATHEMATICAL LIBRARY

published by

The Mathematical Association of America

Editorial Committee

Paul Zorn, Chair Anneli Lax
St Olaf College *New York University*

Judith E. Broadwin *Jericho High School*
Underwood Dudley *DePauw University*
Edward M. Harris
David Sanchez *San Antonio College*
Michael J. McAsey *Bradley University*
Peter Ungar

The New Mathematical Library (NML) was begun in 1961 by the School Mathematics Study Group to make available to high school students short expository books on various topics not usually covered in the high school syllabus. In three decades the NML has matured into a steadily growing series of some thirty titles of interest not only to the originally intended audience, but to college students and teachers at all levels. Previously published by Random House and L. W. Singer, the NML became a publication series of the Mathematical Association of America (MAA) in 1975. Under the auspices of the MAA the NML will continue to grow and will remain dedicated to its original and expanded purposes.

EPISODES IN NINETEENTH AND TWENTIETH CENTURY EUCLIDEAN GEOMETRY

by

Ross Honsberger
University of Waterloo

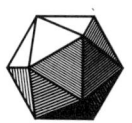

37

THE MATHEMATICAL ASSOCIATION OF AMERICA

© 1995 by
The Mathematical Association of America

All rights reserved under International and Pan American Copyright Conventions
Published in Washington by The Mathematical Association of America

Library of Congress Catalog Card Number 94-079528
Complete Set ISBN 0-88385-600-X
Vol. 37 ISBN 0-88385-639-5

Manufactured in the United States of America

Note to the Reader

This book is one of a series written by professional mathematicians in order to make some important mathematical ideas interesting and understandable to a large audience of high school students and laymen. Most of the volumes in the *New Mathematical Library* cover topics not usually included in the high school curriculum; they vary in difficulty, and, even within a single book, some parts require a greater degree of concentration than others. Thus, while you need little technical knowledge to understand most of these books, you will have to make an intellectual effort.

If you have so far encountered mathematics only in classroom work, you should keep in mind that a book on mathematics cannot be read quickly. Nor must you expect to understand all parts of the book on first reading. You should feel free to skip complicated parts and return to them later; often an argument will be clarified by a subsequent remark. On the other hand, sections containing thoroughly familiar material may be read very quickly.

The best way to learn mathematics is to *do* mathematics, and each book includes problems some of which may require considerable thought. You are urged to acquire the habit of reading with paper and pencil in hand; in this way, mathematics will become increasingly meaningful to you.

The authors and editorial committee are interested in reactions to the books in this series and hope that you will write to: Anneli Lax, Editor, New Mathematical Library, New York University, The Courant Institute of Mathematical Sciences, 251 Mercer Street, New York, N.Y. 10012.

The Editors

NEW MATHEMATICAL LIBRARY

1. Numbers: Rational and Irrational *by Ivan Niven*
2. What is Calculus About? *by W. W. Sawyer*
3. An Introduction to Inequalities *by E. F. Beckenbach and R. Bellman*
4. Geometric Inequalities *by N. D. Kazarinoff*
5. The Contest Problem Book I Annual h.s. math. exams, 1950–1960. Compiled and with solutions *by Charles T. Salkind*
6. The Lore of Large Numbers *by P. J. Davis*
7. Uses of Infinity *by Leo Zippin*
8. Geometric Transformations I *by I. M. Yaglom, translated by A. Shields*
9. Continued Fractions *by Carl D. Olds*
10. Replaced by NML-34
11. } Hungarian Problem Books I and II, Based on the Eötvös
12. } Competitions 1894–1905 and 1906–1928, *translated by E. Rapaport*
13. Episodes from the Early History of Mathematics *by A. Aaboe*
14. Groups and Their Graphs *by E. Grossman and W. Magnus*
15. The Mathematics of Choice *by Ivan Niven*
16. From Pythagoras to Einstein *by K. O. Friedrichs*
17. The Contest Problem Book II Annual h.s. math. exams 1961–1965. Compiled and with solutions *by Charles T. Salkind*
18. First Concepts of Topology *by W. G. Chinn and N. E. Steenrod*
19. Geometry Revisited *by H. S. M. Coxeter and S. L. Greitzer*
20. Invitation to Number Theory *by Oystein Ore*
21. Geometric Transformations II *by I. M. Yaglom, translated by A. Shields*
22. Elementary Cryptanalysis—A Mathematical Approach *by A. Sinkov*
23. Ingenuity in Mathematics *by Ross Honsberger*
24. Geometric Transformations III *by I. M. Yaglom, translated by A. Shenitzer*
25. The Contest Problem Book III Annual h.s. math. exams. 1966–1972. Compiled and with solutions *by C. T. Salkind and J. M. Earl*
26. Mathematical Methods in Science *by George Polya*
27. International Mathematical Olympiads 1959–1977. Compiled and with solutions *by S. L. Greitzer*
28. The Mathematics of Games and Gambling *by Edward W. Packel*
29. The Contest Problem Book IV Annual h.s. math. exams. 1973–1982. Compiled and with solutions *by R. A. Artino, A. M. Gaglione, and N. Shell*
30. The Role of Mathematics in Science *by M. M. Schiffer and L. Bowden*
31. International Mathematical Olympiads 1978–1985 and forty supplementary problems. Compiled and with solutions *by Murray S.Klamkin*
32. Riddles of the Sphinx *by Martin Gardner*
33. U.S.A. Mathematical Olympiads 1972–1986. Compiled and with solutions *by Murray S. Klamkin*
34. Graphs and Their Uses *by Oystein Ore*. Revised and updated *by Robin J. Wilson*
35. Exploring Mathematics with Your Computer *by Arthur Engel*
36. Game Theory and Strategy *by Philip Straffin*
37. Episodes in Nineteenth and Twentieth Century Euclidean Geometry *by Ross Honsberger*

Other titles in preparation.

Contents

Preface ix
Introduction xi
   1. Cleavers and Splitters 1
   2. The Orthocenter 17
   3. On Triangles 27
   4. On Quadrilaterals 35
   5. A Property of Triangles 43
   6. The Fuhrmann Circle 49
   7. The Symmedian Point 53
   8. The Miquel Theorem 79
   9. The Tucker Circles 87
  10. The Brocard Points 99
  11. The Orthopole 125
  12. On Cevians 137
  13. The Theorem of Menelaus 147
Suggested Reading 155
Solutions to the Exercises 157
Index 173

Preface

It is always gratifying to discover that it is within one's ability to appreciate a mathematics book and to read it with pleasure. I have often dreamt what a joy it would be to get to know some of the elementary gems that are surely present in every branch of mathematics, only to be dismayed by the literature I have been able to find. Undoubtedly the gems are there, but they often lie buried in textbooks or comprehensive reference works. One is frequently left with the unhappy choice of undertaking a prolonged study of the field or giving up the idea altogether. While it takes a knowledgeable scholar to write something out of the ordinary, the dedicated specialist can get carried away with discussions that one comes to appreciate only after long and serious study. Unfortunately, this makes it very difficult for general readers to disentangle the elementary gems of their heart's desire. On top of this, what passes for a proof is often so concise or sketchy that it is readily understandable only to someone who already knows the subject.

I would dearly love to be able to promise that you will find no such frustrations in the present work. What I can promise is a collection of essays that does not attempt to cover a large amount of material, and that each topic has been extricated from the mass of material in which it is usually found and given as elementary and full a treatment as is reasonably possible. There is no sense pretending there is any way around the need to lay foundations for one's proofs, but by selecting from the most accessible topics I hope that many readers will be able to delight in these gems with a minimum of preliminaries.

With one exception, everything that is beyond a high school background is developed as the need arises; the basic properties of the nine-point circle are covered in so many places that their proofs have not been included here. You should be advised, however, that derivations given in one essay are not repeated in a later one. Thus, while sampling from the first few essays might not incur a great loss of continuity, the later essays do make use of concepts and results that have been considered in earlier discussions, and you might have to look in an earlier essay for explanations of an item that is encountered later on.

So much depends on your mood when you take up a book like this. Most assuredly, these essays are not intended to be burdensome studies; on the contrary, although some concentration is required for their appreciation, I hope you will look forward to some relaxing entertainment and enjoy them as you would a beautiful piece of music.

It is a pleasure to thank Paul Zorn and the members of the New Mathematical Library Subcommittee, and Don Albers and Beverly Ruedi, for the excellence of their editorial work and guidance of the book through publication. Particular thanks are due Professors Don Chakerian, University of California at Davis, and Basil Gordon, University of California at Los Angeles, for their reviews of the manuscript, which led to many improvements. Very special thanks are due Anneli Lax for her continuing dedication to the New Mathematical Library Series; her extremely careful reading of the manuscript resulted in many clarifications and improvements in presentation.

Introduction

Mathematics today isn't just one subject, it's dozens of subjects. In the last hundred years, and particularly in the last fifty, our knowledge has increased at an unprecedented rate, and it is only right that our school curriculum has evolved in the light of the new discoveries. It is with some regret, however, that we have had to let go of many beautiful old problems of synthetic geometry.

About 1875, Euclidean geometry experienced a meteoric revival that saw hundreds of marvelous new results published in the next half century. Two outstanding books in this field are Nathan Altschiller Court's *College Geometry* and Roger Johnson's *Advanced Euclidean Geometry*; also, anyone attracted to geometry will be sure to enjoy Coxeter and Greitzer's *Geometry Revisited*.

Synthetic geometry is obviously a chain subject, and although many of the new results are of an elementary nature, they often concern fairly complicated configurations and can only be approached comfortably after substantial preliminary discussion; assuredly any effort directed toward their appreciation is always well rewarded. Fortunately, a good number of these delightful little gems can be enjoyed by recalling just a few well-known theorems and deriving an occasional preliminary result.

Special acknowledgment and deepest thanks are due to
Dr. John Rigby
of the University of Wales College of Cardiff for his invaluable contributions to this book.

CHAPTER ONE

Cleavers and Splitters

1. Cleavers

Archimedes must have been delighted when he discovered his "theorem of the broken chord."

(a) ARCHIMEDES' THEOREM. Let M be the midpoint of the arc ACB on the circumcircle of $\triangle ABC$, and let MD be the perpendicular to the longer of AC and BC, say AC, (Figure 1(a)). Then D bisects the polygonal path ACB, that is,

$$AD = DC + CB.$$

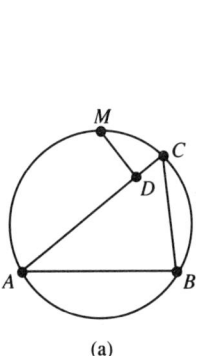 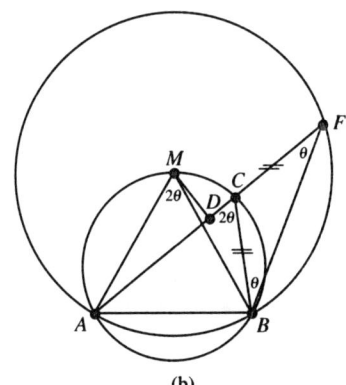

(a) (b)

Figure 1

PROOF. If AC is extended to F so that $CF = CB$, triangle CFB would be isosceles with equal base angles θ at F and B and exterior angle $ACB = 2\theta$ (Figure 1(b)). Now, the circle through A, B, and F would have its center at the point on the perpendicular bisector of AB at which AB subtends an angle which is twice the angle θ it subtends at the point F on the circumference. Clearly, then, the center must be at M, for M is on the perpendicular bisector of AB since it bisects the arc ACB, and the angles $\angle AMB$ and $\angle ACB$ ($= 2\theta$) are equal in the same segment of the given circle. Thus, in the second circle, MD is the perpendicular from the center to the chord AF, making D the *midpoint* of AF, and we have

$$AD = DF = DC + CF = DC + CB,$$

completing the proof. ∎

This argument is very similar to one given by Archimedes himself (compare with the proof on page 150 of Carl Boyer's *History of Mathematics*. Another ingenious proof, by Gregg Patruno, is given in my *More Mathematical Morsels*, Vol. 10, Dolciani Series, page 31.)

By Archimedes' theorem, then, if C' is the midpoint of AB, $C'D$ will have half the perimeter of $\triangle ABC$ on each side of it (Figure 2). Following Dov Avishalom [1], let us call a perimeter-bisecting segment like $C'D$ a *cleaver* when it issues from the *midpoint of a side*.

In Subsection (c) below we shall prove the remarkable fact that

the three cleavers of a triangle always meet in a "cleavance-center."

A nice proof of this can be based on the engaging property that each cleaver is parallel to an angle-bisector of the triangle, a result which follows as an immediate corollary to our proof of Archimedes' theorem.

(b) COROLLARY. *The cleaver $C'D$ is parallel to the bisector of angle C.*

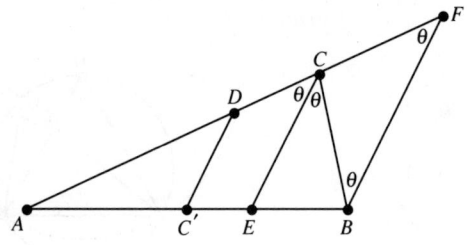

Figure 2

$C'D$ joins the midpoints of two sides of $\triangle ABF$ and is therefore parallel to the third side, BF. But the bisector CE of angle C ($= 2\theta$) in $\triangle ABC$ clearly determines equal corresponding angles at C and F, making CE also parallel to BF, and hence to $C'D$. ∎

(We note in passing that this gives a nice construction for a cleaver.)

(c) THE MEDIAL TRIANGLE. Triangle $A'B'C'$, determined by the feet of the medians, that is, the midpoints of the sides, is called the *medial* triangle of $\triangle ABC$, and clearly has sides that are parallel to those of the given triangle ABC. Thus, in parallelogram $CB'C'A'$ (Figure 3), the opposite angles at C and C' are equal, and so the cleaver $C'D$, being parallel to the bisector CE of angle C, is therefore the bisector of angle C' in the medial triangle (both CE and $C'D$ are inclined at the same angle to the common direction of CA' and $B'C'$). Similarly, the other two cleavers are also angle bisectors in the medial triangle, and

the cleavance-center is simply the incenter S of the medial triangle

(recall that the incenter of a triangle is the center of the inscribed circle and is accordingly the point of intersection of the three angle bisectors).

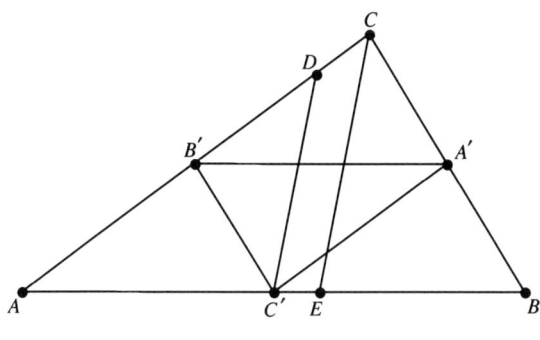

Figure 3

Perhaps it would be more striking to state this in the form:

an angle bisector of the medial triangle of $\triangle ABC$ bisects the perimeter of $\triangle ABC$ (Figure 4).

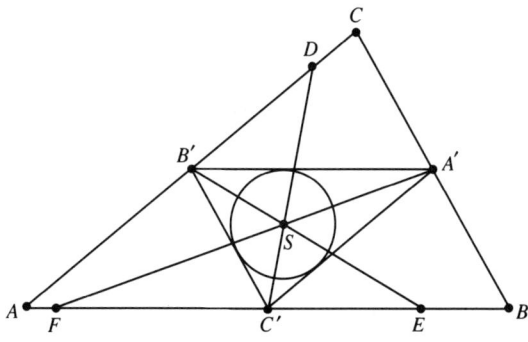

Figure 4

(d) THE SPIEKER CIRCLE. The incircle of the medial triangle is called the *Spieker circle* of the given triangle, in honor of the 19th-century German geometer Theodor Spieker. We have just shown, then, that the cleavance-center of a triangle ABC is the center S of its Spieker circle.

EPISODES

Since each cleaver bisects the perimeter of $\triangle ABC$ and goes through the cleavance-center S, it might not take you completely by surprise to learn that S *is the center of gravity of a homogenous wire frame in the shape of $\triangle ABC$*. Still, this is not obvious, and it has the following nice proof.

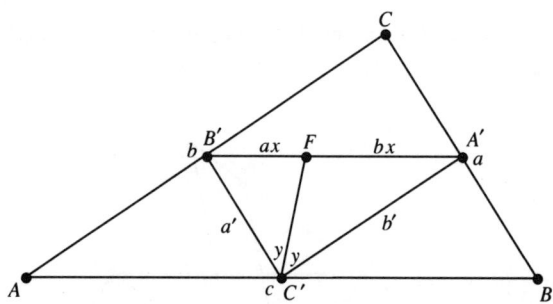

Figure 5

Clearly the frame consists of three wires of lengths a, b, and c. Since the midpoint of a homogenous wire is its center of gravity, the wire frame is equivalent to a system of masses of magnitudes a, b, c, suspended, respectively, at the vertices A', B', C' of the medial triangle. Now the bisector $C'F$ of angle C' in the medial triangle divides the opposite side $B'A'$ in the ratio of the sides about angle C':

$$\frac{B'F}{FA'} = \frac{B'C'}{C'A'} = \frac{a'}{b'}.$$

But the sides of the medial triangle are not only parallel to the sides of $\triangle ABC$, they are also half as long. Hence

$$\frac{B'F}{FA'} = \frac{a'}{b'} = \frac{a/2}{b/2} = \frac{a}{b},$$

and for some real number x we have

$$B'F = ax \quad \text{and} \quad FA' = bx.$$

Consequently the moments about F of the masses at B' and A' are equal,

$$b(ax) = a(bx),$$

and we conclude that the whole system is equivalent to a mass of $(b + a)$ at F and a mass of c at C'. The center of gravity, then, must lie somewhere on the cleaver $C'F$. Lying similarly on each of the other two cleavers, the center of gravity of the system is indeed the cleavance-center S of $\triangle ABC$. ∎

2. Splitters

We have called a perimeter-bisecting segment through the midpoint of a side a cleaver; let us call such a segment which issues from a **vertex** a *splitter*. Thus a triangle ABC also has three splitters and, as you might expect, they meet in a point M called the *splitting-center*.

If the excircle opposite A touches the sides of $\triangle ABC$ at D, E, and F (Figure 6), then clearly

$$BE = BD \quad \text{and} \quad DC = CF,$$

revealing that the perimeter of $\triangle ABC$ can be cut at D and straightened out to lie along the tangents AE and AF. But these tangents are equal, and we have

$$AE = AB + BD = \text{the semiperimeter } s \text{ of } \triangle ABC,$$

making AD a splitter. Thus the concurrence of the splitters is equivalent to the claim that the three segments, each from a vertex to the point on the opposite side where it is touched by the appropriate excircle, are concurrent. But this point of concurrence has been known for a long time and is called the Nagel point M of the triangle.

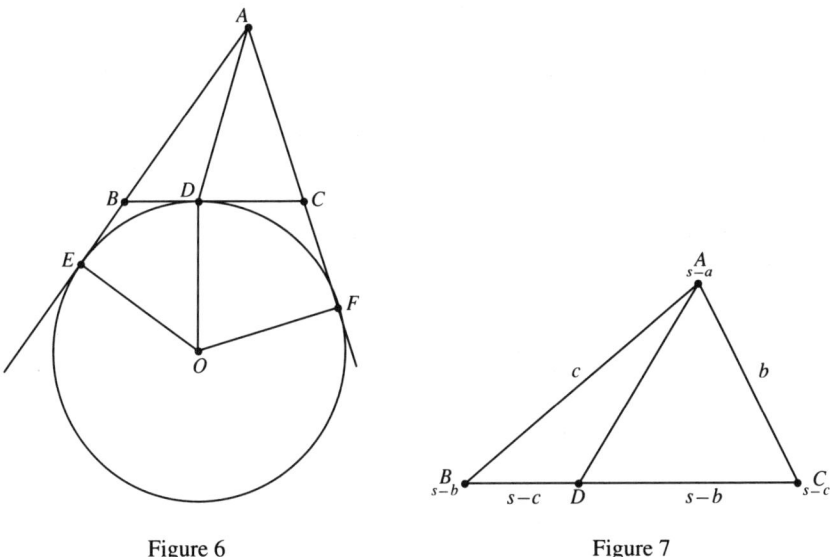

Figure 6   Figure 7

(a) THE NAGEL POINT. The existence of the Nagel point follows very easily from the consideration of a system of masses of magnitudes $s-a$, $s-b$, and $s-c$, suspended respectively at the vertices A, B, and C (Figure 7). Accordingly, let D be the point of contact on BC of the excircle opposite A. Then, as we saw above,

AD is a splitter, and the semiperimeter s is given by

$$s = AB + BD = c + BD.$$

Hence we have

$$BD = s - c.$$

Similarly, $DC = s - b$, and the moments about D of the masses at B and C are equal:

$$(s - b)(s - c) = (s - c)(s - b).$$

Thus D is the center of gravity of the two masses at B and C, and the center of gravity M of the entire system must lie somewhere along AD. Similarly, M lies on each of the other two segments from a vertex to the point of contact on the opposite side of the appropriate excircle, and we conclude that these three segments are indeed concurrent at the center of gravity. This establishes the Nagel point M and hence also the concurrence of the splitters. ∎

3. The Nine-point Circle

The discovery that the circumcircle of the medial triangle $A'B'C'$ passes through not only the three midpoints of the sides but also through six other notable points of $\triangle ABC$ is one of the glories of modern synthetic geometry. It is well known that the altitudes of a triangle concur at the orthocenter H (see page 19 for a proof), and it is customary to call the midpoints P_1, P_2, P_3 of the segments which join H to the vertices the *Euler points* of the triangle (Figure 8). It is remarkable that, for every triangle, there is a nine-point circle which passes through the three midpoints of the sides, the three feet of the altitudes, and the three Euler points. A great deal has been written about this famous circle and we will note in passing only a few of its properties.

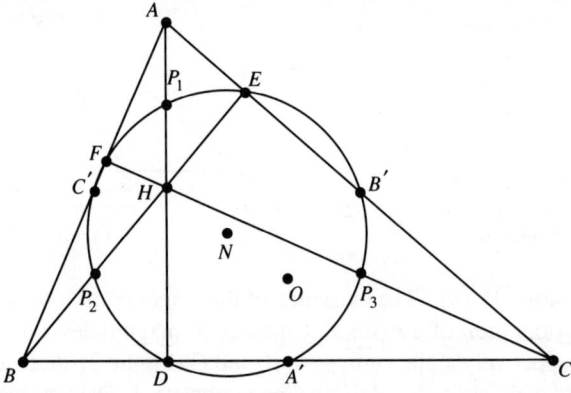

Figure 8

It is common practice to use the letters H, G, O, I, and N to denote, respectively, the orthocenter, centroid, circumcenter, incenter, and the center of the nine-point circle (Figure 8). Recall that

- the orthocenter H is the point of intersection of the three altitudes,
- the centroid G is the point of intersection of the three medians,
- the circumcenter O is the center of the circle through the vertices and is therefore the point of intersection of the perpendicular bisectors of the three sides.

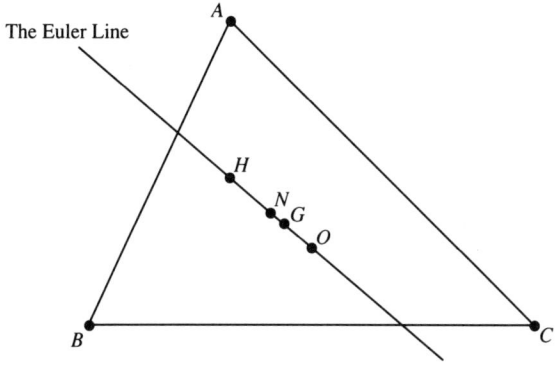

Figure 9

Now, it happens that H, G, and O are always collinear, and the line they determine is known as the *Euler line* (Figure 9). Besides that, G trisects the segment HO so that $HG = 2 \cdot GO$. Now N enters the picture, and not only does N also lie on the Euler line, but in fact is the midpoint of HO. It follows, then, that

$$OG = \frac{1}{3}HO, \quad ON = \frac{1}{2}HO, \quad \text{and} \quad NG = \frac{1}{6}HO.$$

4. The Nagel Point M and the Spieker Circle

As above, let S denote the center of the Spieker circle. Then, remarkably, we have the parallel results that

- I, G, and M are always collinear,
- G trisects IM so that $GM = 2 \cdot IG$, and
- S is the midpoint of IM (Figure 10).

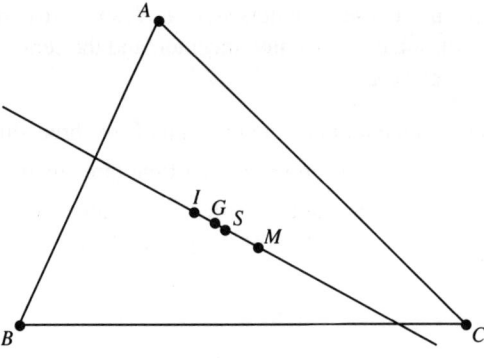

Figure 10

(a) TO PROVE THAT G TRISECTS IM. The centroid G divides the median AA' in the ratio 2 : 1 (Figure 11). Now let IG be extended twice its length to the point L, so that $GL = 2IG$. This makes triangles AGL and $A'GI$ similar (the sides about the equal angles at G are proportional). Accordingly, the angles at A and A' are equal and we conclude that AL is parallel to $A'I$. We shall show that L is the Nagel point by showing that AL, BL, and CL are the three splitters of $\triangle ABC$. We need only consider the typical case of AL.

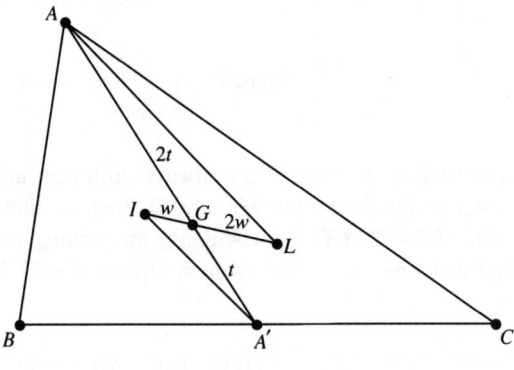

Figure 11

To this end, extend AL to meet BC at E, and let perpendiculars AD and IF be drawn to BC (Figure 12). We need to show that $AB + BE = s$, the semiperimeter, that is, that

$$BE = s - AB = s - c.$$

Accordingly, let us calculate the lengths of BD and DE and show that their sum BE is indeed $s - c$.

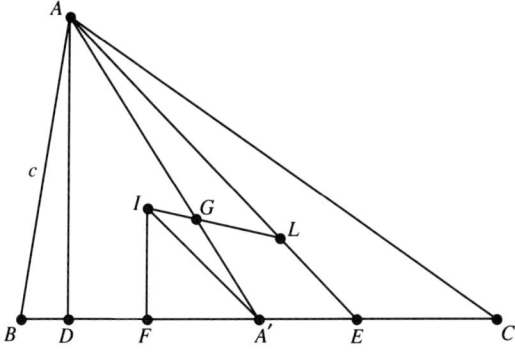

Figure 12

(i) THE LENGTH OF DE. Because $AE \parallel IA'$, the triangles ADE and IFA' are similar, and we have

$$\frac{DE}{FA'} = \frac{AD}{IF}.$$

Now, IF is just the inradius of $\triangle ABC$ and AD is an altitude, and it is well known that the area $\triangle$ of triangle ABC is given by each of the formulas

$$\triangle = rs \quad \text{and} \quad \triangle = \frac{1}{2}a(AD).$$

Thus

$$rs = \frac{1}{2}a(AD),$$

and we have

$$\frac{AD}{IF} = \frac{AD}{r} = \frac{2s}{a}.$$

As a result,

$$\frac{DE}{FA'} = \frac{2s}{a}, \quad \text{and so} \quad DE = \frac{2s}{a}FA'.$$

To FIND FA'. From Figure 13, it is clear that $x + y + z = s$, so $BF = x = s - b$. Hence the length of the tangent BF to the incircle is simply $s - b$. Thus, in Figure 12,

$$FA' = BA' - BF = \frac{1}{2}a - (s - b) = \frac{1}{2}a - s + b$$

$$= \frac{1}{2}a - \frac{1}{2}(a + b + c) + b = \frac{b - c}{2}.$$

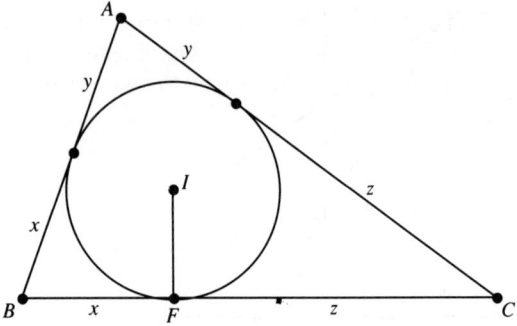

Figure 13

Therefore

$$DE = \left(\frac{2s}{a}\right) FA' = \left(\frac{2s}{a}\right)\left(\frac{b-c}{2}\right) = \frac{s(b-c)}{a}.$$

(ii) THE LENGTH OF BD. Now, from $\triangle ABD$, we have $BD = c\cos B$, and from the law of cosines, we have

$$b^2 = a^2 + c^2 - 2a(c\cos B).$$

Hence

$$BD = c\cos B = \frac{a^2 + c^2 - b^2}{2a}.$$

(iii) THEIR SUM. Finally,

$$\begin{aligned}
BE &= BD + DE \\
&= \frac{a^2 + c^2 - b^2}{2a} + \frac{s(b-c)}{a} \\
&= \frac{a^2 + c^2 - b^2 + 2s(b-c)}{2a} \\
&= \frac{a^2 + c^2 - b^2 + (a+b+c)(b-c)}{2a} \\
&= \frac{a^2 + ab - ac}{2a} \\
&= \frac{a+b-c}{2} \\
&= \frac{1}{2}(a+b+c) - c \\
&= s - c,
\end{aligned}$$

as desired. Thus $L = M$ and G trisects IM so that $IG = \frac{1}{3}IM$. ∎

(b) To Prove That S Bisects IM. Again, the proof brings together some very basic and useful geometry (but this time it's quite elegant). A brief introduction to dilatations is given in the appendix (on page 15) for those who are not familiar with this transformation.

Since the centroid G trisects each median, the dilatation $G(-\frac{1}{2})$ transforms the given triangle ABC into its medial triangle $A'B'C'$ (Figure 14). Under this transformation, the incircle of ABC is carried to the incircle of $A'B'C'$, which is just the Spieker circle of ABC. Because the image of a circle under a dilatation is also a circle and the center of a circle is carried to the center of its image-circle, our transformation takes I to S. In order for $G(-\frac{1}{2})$ to do this, I, G, and S must be collinear with $IG = 2GS$. Since $IG = \frac{1}{3}IM$, then $GS = \frac{1}{6}IM$, and adding, we obtain

$$IS = \frac{1}{2}IM.\ \blacksquare$$

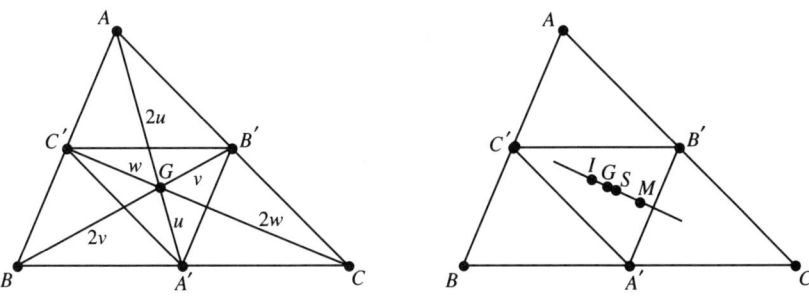

Figure 14

(c) Two Additional Properties.

(i) The dilatation $G(-\frac{1}{2})$ not only carries $\triangle ABC$ onto the medial triangle $A'B'C'$, but because M, G, and I are collinear and $MG = 2GI$, it also takes M to I. Thus the medial triangle $A'B'C'$ with the point I inside it is merely a half-size copy of

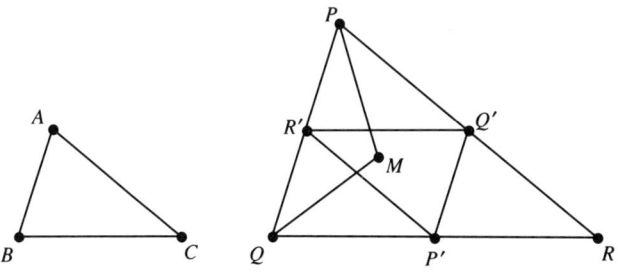

Figure 14(a)

△*ABC* and the point *M* inside it. It follows that

the incenter of a triangle is the Nagel point of its medial triangle.

This gives a nice construction for the Nagel point (Figure 14a): in a triangle *PQR* with sides that are twice as long as those of a given triangle *ABC*, two angle bisectors give the Nagel point *M* of the medial triangle $P'Q'R'$, which is just a copy of the given triangle *ABC*.

(ii) We conclude this little excursion with the following engaging property of the Nagel point, a property that we shall find is unexpectedly easy to prove.

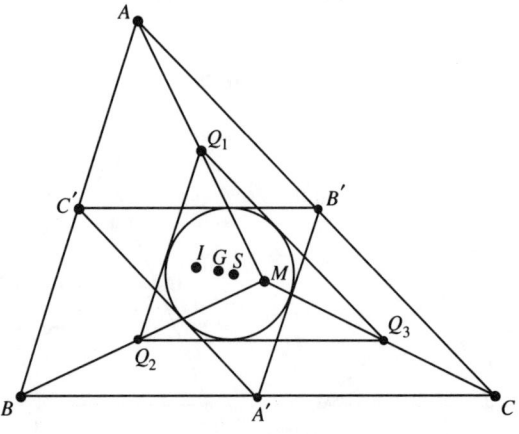

Figure 15

If Q_1, Q_2, and Q_3 are the midpoints of the segments which join the Nagel point *M* to the vertices *A, B, C* (shades of the Euler points), then

△$Q_1Q_2Q_3$ is congruent to the medial triangle, and moreover *each side of* △$Q_1Q_2Q_3$ *is tangent to the Spieker circle*. Thus the Spieker circle is the *common incircle* of the two congruent triangles $Q_1Q_2Q_3$ and $A'B'C'$.

The compound transformation *T* consisting of the dilatation *M*(2) followed by the dilatation $G(-\frac{1}{2})$ first takes △$Q_1Q_2Q_3$ to △*ABC*, and then further takes △*ABC* to △$A'B'C'$. Since *M*(2) doubles the lengths of the sides of a triangle and $G(-\frac{1}{2})$ halves them, the lengths of the sides of the resulting triangle $A'B'C'$ are the same as those of △$Q_1Q_2Q_3$ and the triangles are congruent.

Now, *T* takes the incircle △$Q_1Q_2Q_3$ to the incircle of △$A'B'C'$ and, in particular, it takes the incenter I_q of △$Q_1Q_2Q_3$ to the incenter *S* of △$A'B'C'$. Next, let us consider where *S* is taken by *T*. Because *S* bisects *IM*, the dilatation *M*(2) takes *S* to *I*; but because $IG = 2GS$, the second dilatation $G(-\frac{1}{2})$ carries *I* back to *S*. Thus *S* is its *own image* under *T*. Recalling that *T* takes the incenter I_q to *S*, it must be that I_q is S itself. That is to say, the two incircles have the same center, and since they have the same radius (in congruent triangles), they must be the same circle. ∎

Comment

Any segment which runs from a vertex to the opposite side of a triangle is called a *cevian* in honor of the seventeenth century Italian engineer Giovanni Ceva.

Now, clearly a dilatation takes lines which are concurrent at a point P into lines that are concurrent at the image of P. Moreover, because dilatations take lines into *parallel* lines, the dilatation $G(-\frac{1}{2})$ would carry a cevian of $\triangle ABC$ into a parallel segment through the corresponding vertex of the medial triangle. Therefore,

> given three cevians that are concurrent at a point P inside $\triangle ABC$, the corresponding parallels through the vertices A', B', and C' of the medial triangles are also concurrent (at the image of P under $G(-\frac{1}{2})$).

Cases of special interest include

(a) Parallels to the angle bisectors of a triangle give the cleavers (providing a quick proof of their concurrence).
(b) Parallels to the medians are just the medians themselves, implying a triangle and its medial triangle have the same centroid.
(c) Parallels to the splitters are concurrent.

Exercise Set 1

1.1 (Refer to the item immediately preceding the comment above.) Since S is invariant under the compound transformation T, and S is the midpoint of IM, one might wonder whether T is equivalent to a half-turn about S (so that Q_1, S, A' are collinear, etc.). Prove or disprove this conjecture.
1.2 Suppose X is the point of contact of the incircle of $\triangle ABC$ on the side BC. Prove that the incircles of triangles ABX and ACX touch AX at the same point P. Prove the corresponding property for excircles. See Figure 16.
1.3 (An alternative proof that S bisects IM)
   (a) Prove that the center of gravity of masses a, b, c, suspended respectively at A, B, C, is the incenter I of $\triangle ABC$.
   (b) Prove that the center of gravity of masses $2s - 2a, 2s - 2b, 2s - 2c$, suspended respectively at A, B, C, is the Nagel point M of $\triangle ABC$.
   (c) Prove that the center of gravity of masses $b + c, c + a, a + b$, suspended respectively at A, B, C, is the incenter S of the medial triangle of $\triangle ABC$. Thus prove that S is the midpoint of IM.
1.4 (An alternative proof that G trisects IM)
   (a) Prove that the center of gravity of masses a, b, c, suspended respectively at A, B, C, is the incenter I of $\triangle ABC$. (This is the same as Exercise 1.3(a).)

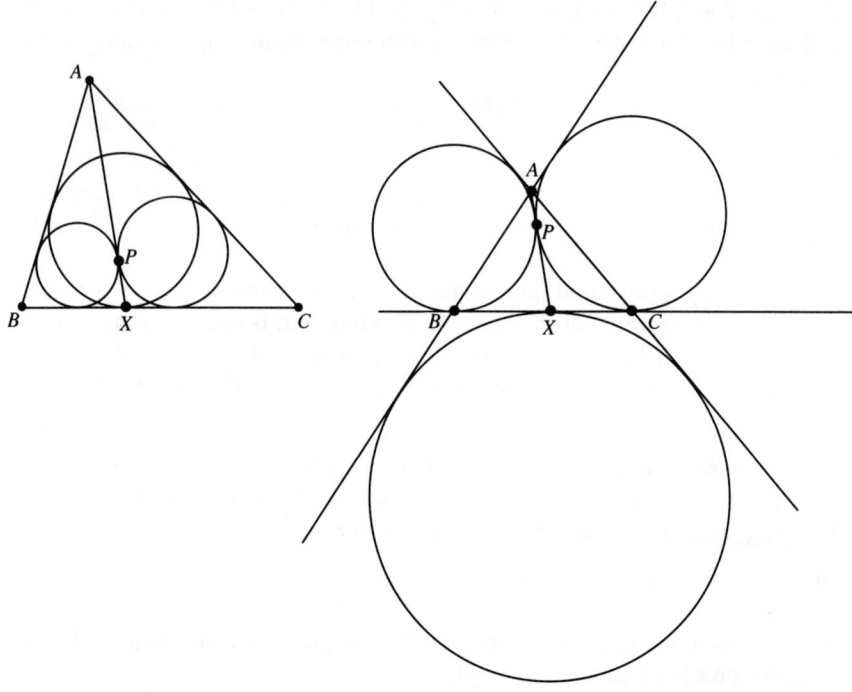

Figure 16

(b) Prove that the center of gravity of masses $s - a, s - b, s - c$, suspended respectively at A, B, C, is the Nagel point M of $\triangle ABC$. (This follows immediately from Exercise 1.3(b).)

(c) Observe that the system in (b) has *half* the total mass of the system in (a), and that a system of equal masses at A, B, C would have center of mass at the centroid G. Prove that G trisects IM.

References

[1] Dov Avishalom, *Mathematics Magazine*, 1963, 60–62.
[2] Roger Johnson, *Advanced Euclidean Geometry*, 225–227.

APPENDIX

Dilatations

The dilatation $O(\mu)$ carries a point P to an image P', collinear with O and P, such that $OP' = |\mu| \cdot OP$, with the stipulation that, if μ is negative, P' and P are on opposite sides of O. Thus $O(-\frac{1}{2})$ sends a point P to the point P' on the line OP a distance half as far on the other side of O:

$O\left(-\dfrac{1}{2}\right)$: P •————2————• O ——1—— • P'

$O\left(\dfrac{1}{2}\right)$: P •———— P' •———— • O

Two Important Properties

(a) SHAPES ARE UNALTERED. While the linear dimensions of a figure are multiplied by $|\mu|$, and hence the area becomes μ^2 times as great, the shape is not changed; the image is *similar* to the original figure, and accordingly a dilatation is said to be a *similarity* transformation. This is established simply by showing that $O(\mu)$ carries a segment AB into a *parallel* segment $A'B'$ which is $|\mu|$ times as long.

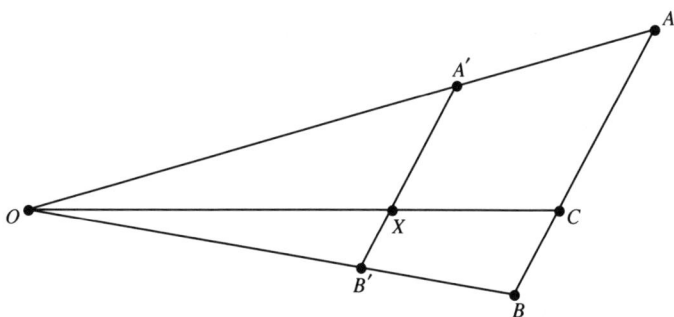

Figure 17

Let $O(\mu)$ take A to A' and B to B'. Let C be an arbitrary point on AB, and let OC cross $A'B'$ at X (Figure 17). Because

$$\frac{OA'}{OA} = \frac{OB'}{OB}(= |\mu|),$$

$A'B'$ is parallel to AB, and from this it follows that triangles $OA'B'$ and OAB are similar, with corresponding sides in the ratio $|\mu| : 1$. Hence $A'B'$ is parallel to AB and $|\mu|$ times as long. It remains to shows that $A'B'$ is in fact the image of AB.

The parallel lines also give $\triangle OA'X$ similar to $\triangle OAC$, and we have

$$\frac{OX}{OC} = \frac{OA'}{OA} = |\mu|.$$

Thus $OX = |\mu| OC$, showing that X is the image of C. We conclude that every point on AB has its image on $A'B'$, and conversely, as desired. ∎

(b) CIRCLES GO TO CIRCLES AND CENTERS TO CENTERS.

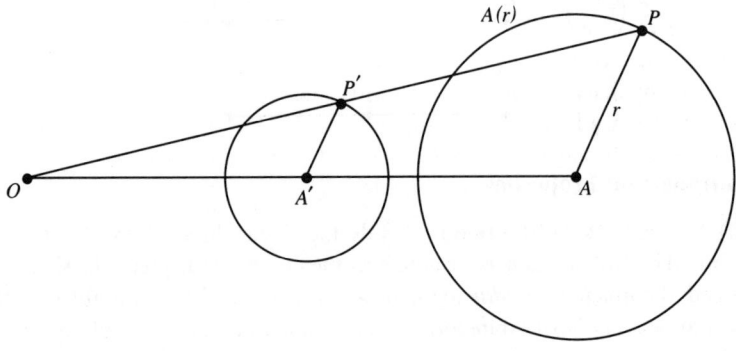

Figure 18

We have seen that $O(\mu)$ takes AP to a parallel segment $A'P'$ that is $|\mu|$ times as long as AP. Thus, if P moves so that AP has a fixed length r, then P' moves so that $A'P'$ has the constant length $|\mu| r$, implying that the image of the circle with center A and radius r is the circle with center A' and radius $|\mu| r$.

CHAPTER TWO

The Orthocenter

1. Let's begin with a neat argument which shows that *the altitudes of a triangle ABC are concurrent*. Following the initial step we took in the previous essay in showing that G trisects IM, let the circumcenter O be joined to the centroid G and extended twice as far to a point H to make $GH = 2 \cdot OG$ (Figure 19). Now, G divides the median AA' in the ratio of $2:1$. Thus in triangles AHG and $A'OG$ the arms of the vertically opposite angles at G are proportional, making the triangles similar. Hence the angles in these triangles at A and A' are equal, so AH and OA' are parallel. But OA' is perpendicular to BC, and therefore so is AH. That is to say, the point H lies on the altitude from A. But vertex A isn't special; similarly, H lies on the altitudes from B and C, and we conclude that the altitudes are indeed concurrent at H, a point which is known as the *orthocenter* of $\triangle ABC$. ∎

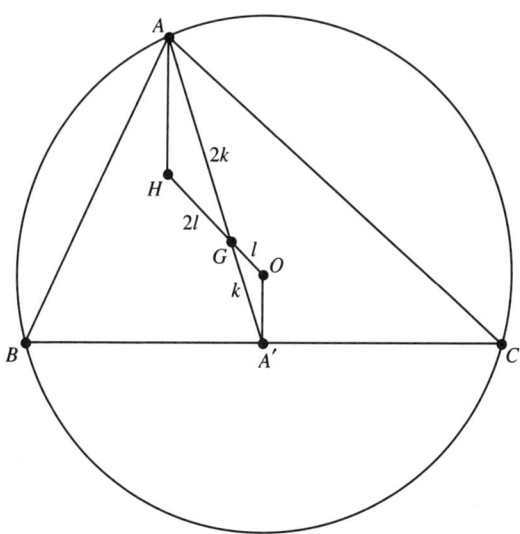

Figure 19

2. A property that might come as a mild surprise is that

if an altitude is extended to the circumcircle, the extension has the same length as the part between the orthocenter and the foot of the altitude; e.g. in Figure 20,

$$DD' = HD.$$

Since the location of the orthocenter is not known in readily usable terms, you might feel the need to brace yourself for an awkward proof. On the contrary, however, there could hardly be anything simpler than the following argument.

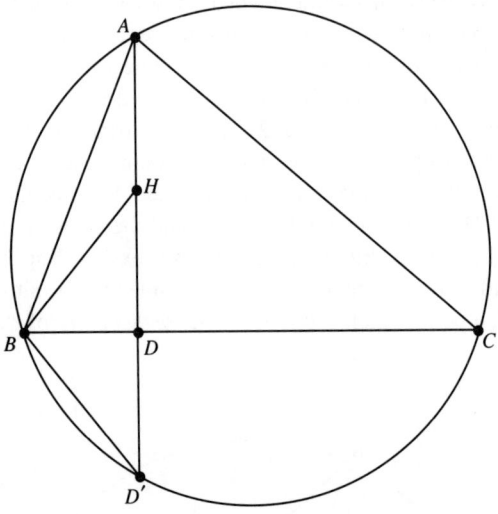

Figure 20

The angles at D' and C are equal since both subtend the arc $\overline{AB}$. But $\angle BHD'$ is also equal to $\angle C$; this follows from the keen observation that the arms of the two angles are respectively perpendicular:

$$HD \perp BC \quad \text{and} \quad \text{(altitude) } BH \perp AC.$$

Thus $\triangle BHD'$ is isosceles, so the altitude BD to the base HD' bisects it. ∎

This result has many consequences. Let us note, in passing, just three of its immediate corollaries.

(a) Reflecting the figure in the side BC would carry D' to H, while leaving B and C fixed. That is to say, the *circle* around $\triangle BCD'$, which is just the circumcircle of $\triangle ABC$ itself, would be carried to the circumcircle of $\triangle HBC$ (Figure 21). Similar reflections in the other sides, then, give us the result that

the circumcircles of the four triangles ABC, HBC, HAC, and HAB all have the same radius. ∎

THE ORTHOCENTER

(b) *The product of a whole altitude AD of a triangle and the part of this altitude between its foot and the orthocenter HD is equal to the product of the two segments into which the altitude divides the side.* In Figure 21

$$BD \cdot DC = AD \cdot HD.$$

This is an immediate consequence of the theorem that the product of the parts of intersecting chords of a circle are equal (proved by means of similar triangles) and Property 2 above which states that $DD' = HD$. ∎

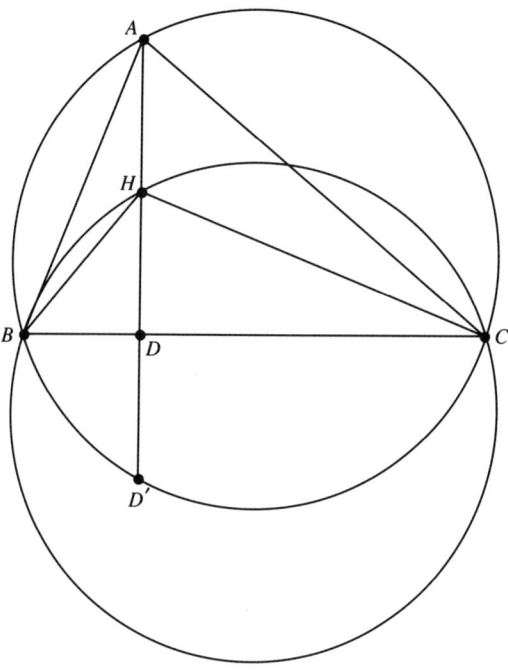

Figure 21

The well known fact that the altitude to the hypotenuse of a right triangle divides the hypotenuse into two parts whose product is the square of the altitude,

$$BD \cdot DC = AD^2$$

(Figure 22), may be considered a special case of property (b). This can be seen, for example, by letting chord BC be the diameter of the circumcircle of $\triangle ABC$ (Figure 21), so that there is a right angle at A, H and A coincide, $HD = AD$, and so $BD \cdot DC = AD \cdot HD = AD^2$ (Figure 22). ∎

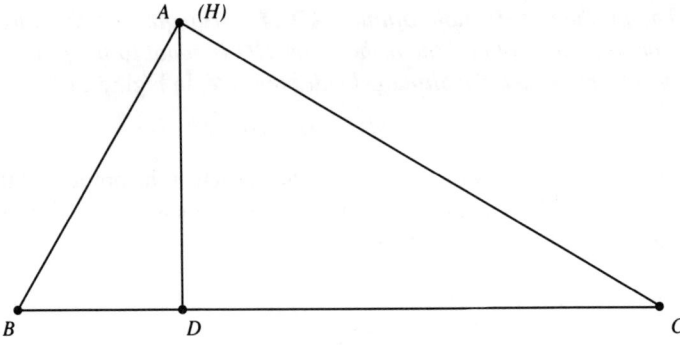

Figure 22

(c) Again, chords AD' and BE' intersect at H (Figure 23) to give $AH \cdot HD' = BH \cdot HE'$, that is, $AH \cdot 2HD = BH \cdot 2HE$, and therefore $AH \cdot HD = BH \cdot HE$. Similarly, $CH \cdot HF$ has the same value, and we have the general result that

the product of the parts into which the orthocenter divides an altitude is the same for all three altitudes. ∎

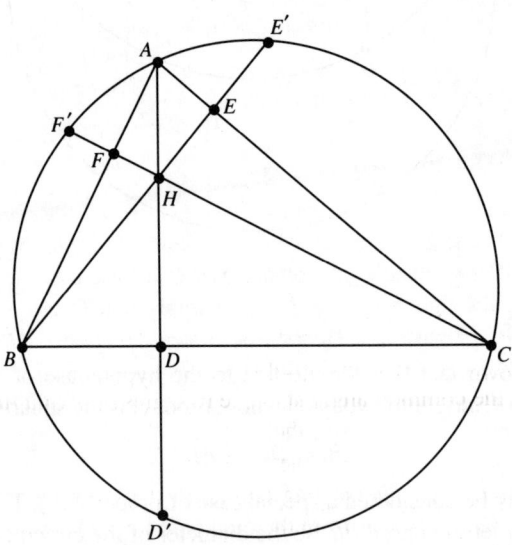

Figure 23

THE ORTHOCENTER

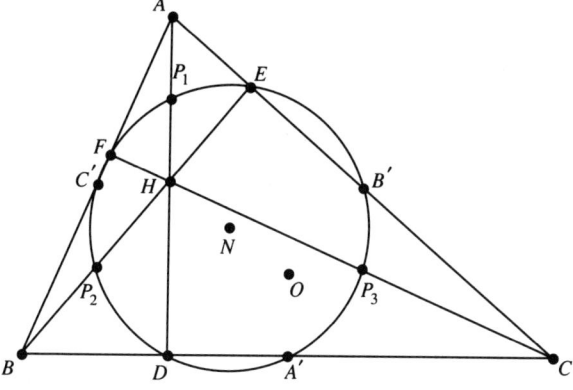

Figure 24

We note that this property is also intimately connected with the nine-point circle (Figure 24). Since the Euler points bisect the segments AH, BH, and CH, the equal products of the segments along intersecting chords in the nine-point circle give

$$P_1H \cdot HD = P_2H \cdot HE = P_3H \cdot HF,$$

that is,

$$\frac{1}{2}AH \cdot HD = \frac{1}{2}BH \cdot HE = \frac{1}{2}CH \cdot HF,$$

from which the property follows immediately.

3. The Orthic Triangle

(a) The triangle DEF, determined by the feet of the altitudes, is called *the orthic triangle* of $\triangle ABC$. Perhaps the most evident property of the orthic triangle is the fact that it cuts off at each vertex of ABC a little triangle that is similar to $\triangle ABC$ itself: the right angles at F and E imply that F and E lie on a circle with diameter BC (Figure 25). Hence, for an acute angled $\triangle ABC$, the interior angle $\angle FBC$ and the exterior angle $\angle AEF$ are equal, for each is the supplement of $\angle FEC$. With the common angle at A, the triangles AEF and ABC have equal corresponding angles; similarly for the little triangles at the other vertices.

We might note in passing that the property (c) in Section 2 follows from the observations that the intersecting chords in the circle around $FBCE$ give $BH \cdot HE = CH \cdot HF$, and that from the circle around $FDCA$ we similarly have $CH \cdot HF = AH \cdot HD$.

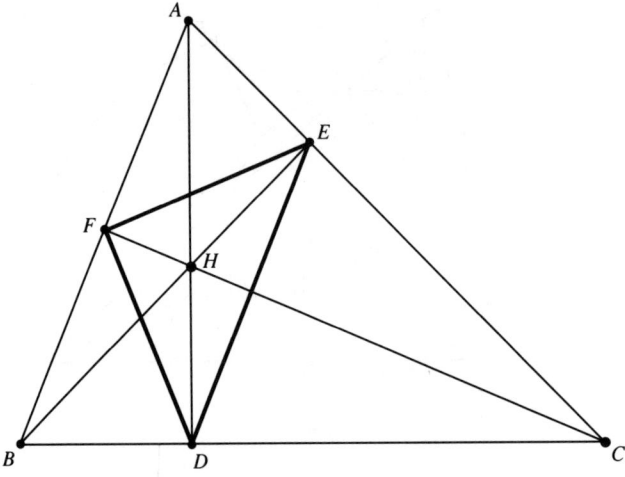

Figure 25

(b) Now let's apply the basic result $DD' = HD$ to establish the striking property that

The radius of the circumcircle of $\triangle ABC$ to a vertex is always perpendicular to the corresponding side of the orthic triangle: in Figure 26,

$$OA \perp EF, \quad OB \perp FD, \quad OC \perp DE.$$

The proof is quite pretty, although there is an extremely simple proof based on the first exercise in the essay on the Tucker circles, page 98.

Because $HE = EE'$ and $HF = FF'$, EF joins the midpoints of two sides of $\triangle HF'E'$, and we have $EF \parallel E'F'$.

Next we establish the interesting preliminary property that *the vertex A is the midpoint of the arc $F'AE'$.* (Similarly, B and C bisect the arcs $F'BD'$ and $D'CE'$). Clearly the arcs $F'A$ and AE' are equal if and only if they subtend equal angles $z = \angle F'CA$ and $w = \angle ABE'$ (Figure 26).

Clearly

$$z = 90° - A \quad \text{in right triangle } AFC;$$

and similarly

$$w = 90° - A \quad \text{in right triangle } ABE.$$

Hence $z = w$, as desired.

It follows immediately that $OA \perp E'F'$, for the radius to the midpoint of an arc is the perpendicular bisector of the chord joining its ends; thus OA is also perpendicular to the parallel segment EF. ■

THE ORTHOCENTER

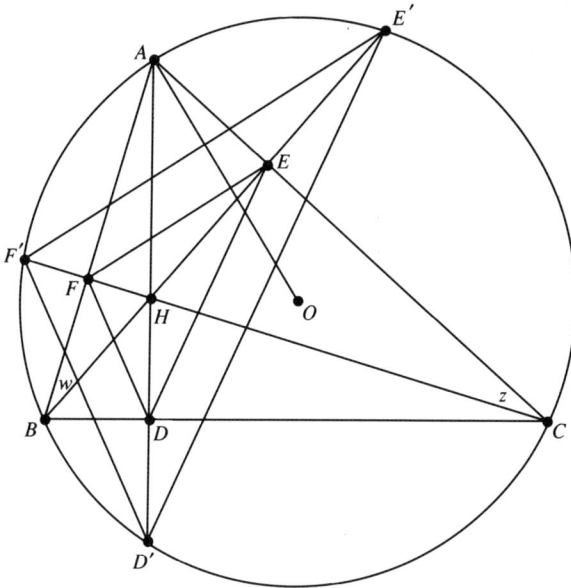

Figure 26

4. Finally, a property that strikes me as most unexpected:

the distance along an altitude from a vertex to the orthocenter is always just twice the distance from the circumcenter to the opposite side of the triangle; e.g. in Figure 27,

$$AH = 2 \cdot OA'.$$

Let the radius BO be extended to the diameter BK. Then KC is perpendicular to BC, so KC is parallel to the altitude AHD.

Also, because BK is a diameter, AK is perpendicular to AB, and hence AK is parallel to the altitude CHF. Thus the opposite sides of $AHCK$ are parallel, making it a parallelogram. Therefore the opposite sides AH and KC are equal. But $KC = 2 \cdot OA'$ from the similar triangles KCB and $OA'B$, so

$$AH = 2 \cdot OA'. \blacksquare$$

While this is a nice proof, we might note that, having established the concurrence of the altitudes by the argument given at the beginning of this essay, this result follows immediately from the fact that corresponding sides of the similar triangles AHG and GOA' are in the ratio of 2 : 1 (as a glance at Figure 19, page 17, confirms). We might also note that, since G trisects both AA' and HO as in Figure 19, the dilatation $G(-\frac{1}{2})$ carries AH into $A'O$, again giving $AH = 2 \cdot OA'$ immediately.

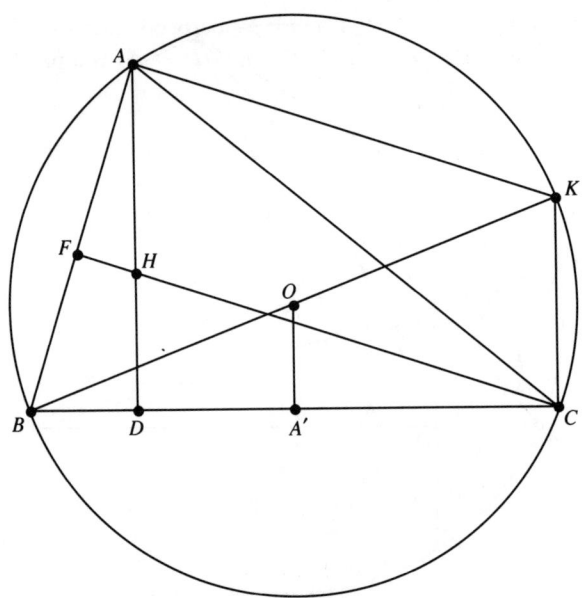

Figure 27

Let us conclude with a quick corollary and a neat application of this result:

COROLLARY. *If two triangles A_1BC and A_2BC are inscribed in a circle on a common base BC, the segment H_1H_2 which joins their orthocenters is equal and parallel to the segment A_1A_2 (Figure 28).*

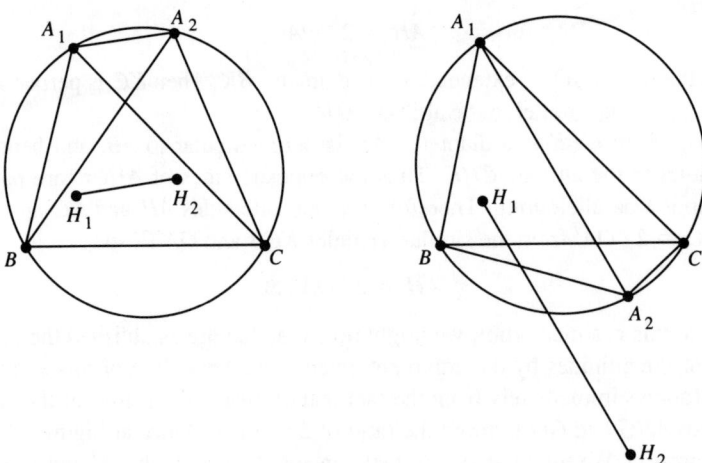

Figure 28

THE ORTHOCENTER

Since each of A_1H_1 and A_2H_2 is not only perpendicular to BC but equal to twice OA', they are equal and parallel. Thus $A_1H_1H_2A_2$ is a parallelogram and the conclusion follows. ∎

APPLICATION. Suppose the orthocenter H of triangle ABC is reflected in vertex C, then C is reflected in the midpoint C' of AB. Prove that the segment joining the images H_1 and C_1 not only passes through the circumcenter O of the triangle but is bisected there (Figure 29).

Let's not construct the figure as described, but simply reflect C in C' to get C_1 and then extend both C_1O and HC to meet at some point Z (Figure 29). We shall show that O is the midpoint of C_1Z and then show that Z is really H_1 by showing that C is the midpoint of HZ.

From the reflection of C in C', we know that C' is the midpoint of CC_1. Because C' is also the midpoint of the side AB, the segment OC' (from the center to the midpoint of a chord) is perpendicular to AB. But because HC is part of an altitude, HCZ is also perpendicular to AB, and we conclude that $C'O$ and CZ are parallel. Thus, in $\triangle C_1CZ$, $C'O$ issues from the midpoint of CC_1 and is parallel to CZ, and hence it bisects the third side C_1Z, making O the midpoint of C_1Z.

We also have that the parallel side CZ is twice as long as $C'O$. But, by the result of this section, we know that HC is twice $C'O$, and hence $HC = CZ$. Thus Z is indeed H_1 and the conclusion follows. ∎

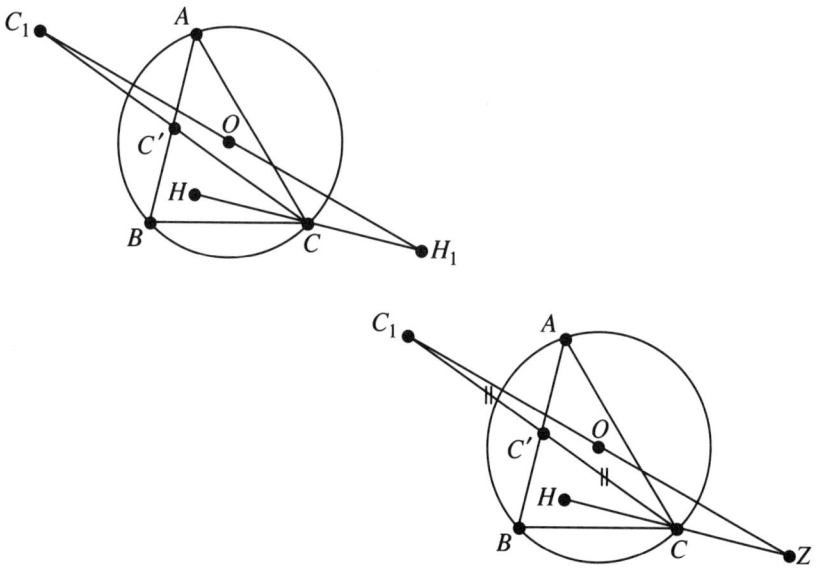

Figure 29

Exercise Set 2

2.1 Through the orthocenter H of $\triangle ABC$ a parallel to AB meets BC at D and a parallel to AC meets BC at E. Perpendiculars to BC at D and E meet AB and AC in D' and E'. Prove that $D'E'$ crosses the circumcircle at the points B' and C' which are diametrically opposite vertices B and C (Figure 30).

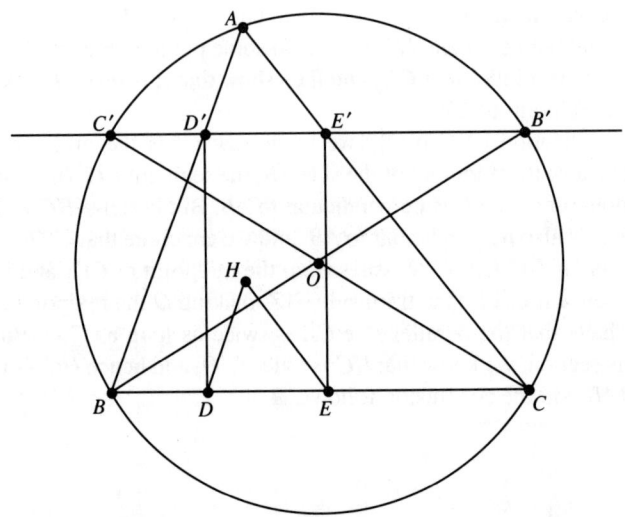

Figure 30

CHAPTER THREE

On Triangles

1. The wave of new properties of a triangle uncovered since the nineteenth-century resurgence of interest in Euclidean geometry is nothing short of a deluge. So many successful investigations have been carried out concerning the concurrences and collinearities of likely and unlikely combinations of lines and circles, parallels, perpendiculars, midpoints, angle-bisectors and the like, that the associated diagrams often grow to be forbiddingly complicated. Of course, with serious study, one's apprehensions soon give way to an appreciation of the underlying order and beauty of these deeper properties. Even though we shall confine ourselves to the simpler discoveries, certain preliminaries are usually necessary to set the stage and gather tools for the proofs. However, I hope that a leisurely pace and the assumption of a minimum of background will make this little episode a readable and enjoyable excursion into this fascinating world of surprising relationships.

2. A Trio of Nested Triangles

We have previously encountered the medial triangle of a given triangle ABC, determined by the midpoints A', B', and C' of its sides. Now let us consider also the triangle around ABC that is determined by the centers I_a, I_b, I_c of its excircles (these centers lie on the bisectors of the exterior angles of $\triangle ABC$). This "excenter triangle," with the given triangle ABC and its medial triangle, form a little nest which always enjoys the following remarkable property (Figure 31):

- from the excenter $\triangle I_a I_b I_c$, pick its circumcenter O_e;
- from $\triangle ABC$ take its orthocenter H,
- and from the medial triangle $A'B'C'$ take its incenter S.

Then the three points H, S, and O_e are collinear, and S is the midpoint of HO_e:

The midpoint of the segment joining the orthocenter of a triangle to the circumcenter of its excenter triangle is the incenter of the medial triangle.

It takes some doing even to imagine how a thing like this would ever be discovered, let alone to come up with a proof for it. Nevertheless, its proof is easy, short

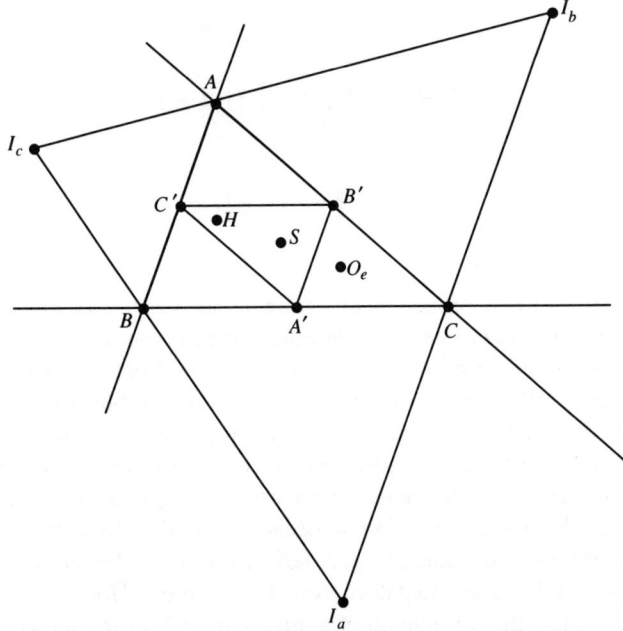

Figure 31

and beautiful, and doesn't tax us with difficult preliminaries. We need only to refresh our memories of the best known property of the Euler line of a triangle, use it to establish an easy result, and then proceed with the concise proof.

Recall that the nine-point circle (whose center we denote by N) of a triangle goes through the three midpoints of the sides, the three feet of the altitudes, and the three Euler points. We have seen that the orthocenter H, the nine-point center N, the centroid G, and the circumcenter O are collinear and that G trisects HO while N bisects it. Let's relate this to our nest of triangles.

Since the excircles are tangent to the sides of $\triangle ABC$, as is the incircle, it is clear that the four centers I, I_a, I_b, and I_c lie on the internal and external angle-bisectors of $\triangle ABC$. But the internal and external angle-bisectors at a vertex are perpendicular (in Figure 32, $2x + 2y = 180°$, so $x + y = 90°$), and we have the rather unexpected result that I_aA, I_bB, and I_cC are the altitudes of $\triangle I_aI_bI_c$, making I the orthocenter of $\triangle I_aI_bI_c$, i.e.,

the incenter I of a triangle is the orthocenter H_e of its excenter triangle.

Thus I and O_e play the roles of orthocenter (H) and circumcenter (O) on the Euler line of $\triangle I_aI_bI_c$.

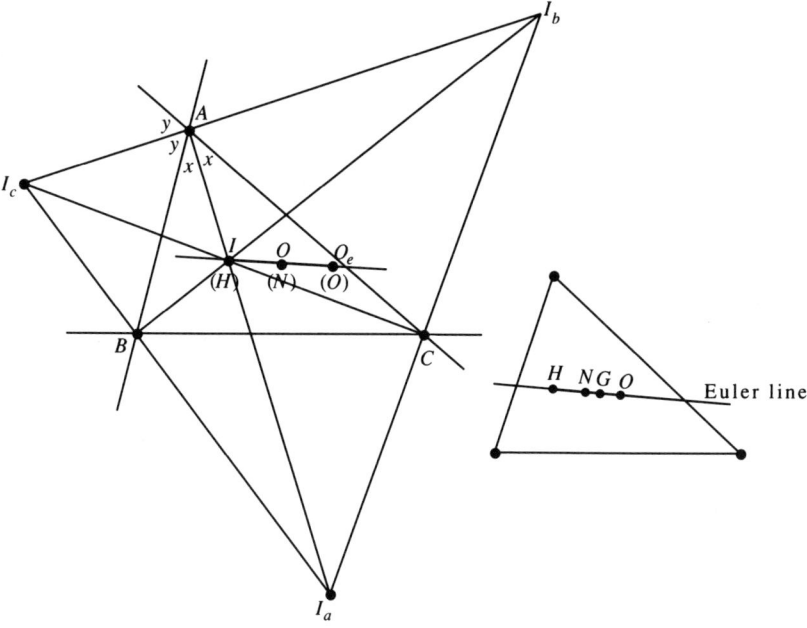

Figure 32

Now, the nine-point circle is the circle through the feet of the altitudes; for $\triangle I_a I_b I_c$, then, these feet are the given vertices A, B, C, themselves, and so the circumcenter O of $\triangle ABC$ is the nine-point center N_e of the excenter triangle. Since N bisects HO on an Euler line, N_e bisects $H_e O_e$ on the Euler line of $\triangle I_a I_b I_c$, and we have that O is the midpoint of IO_e (Figures 32 and 33), i.e., the circumcenter of a triangle bisects the segment which joins its incenter to the circumcenter of its excenter triangle.

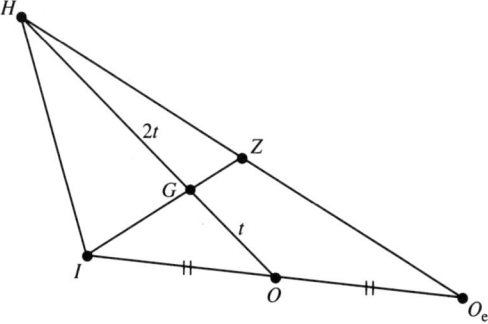

Figure 33

Now to the proof. We can concentrate on the small part of the figure shown in Figure 33, where H is the orthocenter of $\triangle ABC$. Since HO is the Euler line of $\triangle ABC$, its centroid G trisects HO so that $HG : GO = 2 : 1$. Now, because O is the midpoint of IO_e, HO is a median of $\triangle HIO_e$, and since G divides this median in the ratio of $2 : 1$, G must also be the centroid of $\triangle HIO_e$. In this case, IGZ is also a median, making Z the midpoint of HO_e, and it remains only to show that Z is really S, the incenter of the medial triangle of $\triangle ABC$.

To see this, consider the dilatation $G(-\frac{1}{2})$. The centroid G also divides median IZ in the ratio of $2 : 1$, implying that $G(-\frac{1}{2})$ sends I to Z. But, as we have seen on earlier occasions, $G(-\frac{1}{2})$ takes $\triangle ABC$ into its medial triangle $A'B'C'$ and, in doing so, carries the incenter I into the incenter S of the medial triangle. Hence Z is actually the center S, and the proof is complete. ∎

3. An Unlikely Collinearity

Let M be the midpoint of the altitude BE in $\triangle ABC$ and suppose that the excircle opposite B touches AC at Y. Then

the line MY goes through the incenter I!

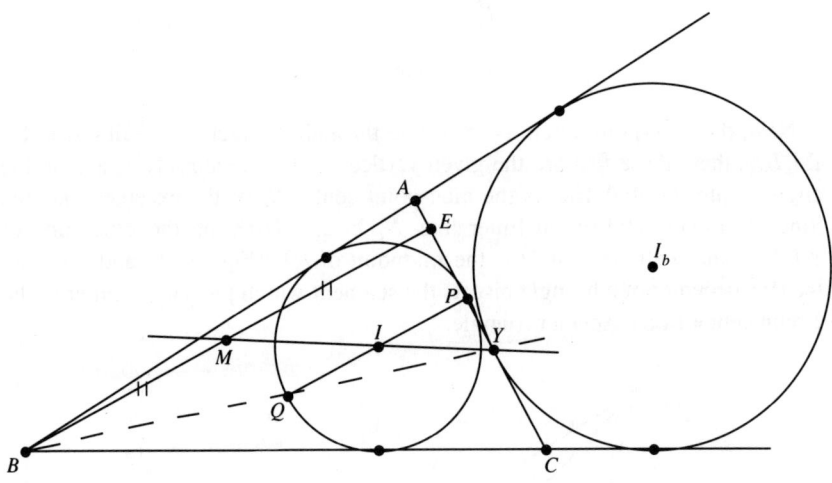

Figure 34

PROOF (John Rigby). Let the incircle of $\triangle ABC$ touch AC at P, and let PQ be the diameter at P (Figure 34). Then PQ is perpendicular to AC, making it parallel to the altitude BE. The fact that M and I are the midpoints of these parallel segments BE and QP implies that M, I, and Y would be collinear if B, Q, and Y were collinear. Indeed, if B, Q, and Y *are* collinear, then, in triangle BYE, YM would be the median to BE and PQ a parallel to BE across the triangle. Since a median bisects every such parallel segment across a triangle (a simple exercise), YM would certainly go through the midpoint I of PQ. Accordingly, let us proceed with the easy proof that B, Q, and Y are in fact collinear, a result that is not unworthy of theorem status on its own.

B, Q, AND Y ARE COLLINEAR. Let XZ be the tangent to the incircle at Q (Figure 35). Since PQ is a diameter, XZ is parallel to the opposite tangent AC. Now, the incircle of $\triangle ABC$ is an excircle of $\triangle BXZ$, and the dilatation $B\left(\frac{BA}{BX}\right)$ takes $\triangle BXZ$ and its excircle (center I) onto $\triangle BAC$ and its excircle (center I_b). In doing so, it also takes the point of contact Q of circle I to the point of contact Y of its image, implying that B, Q, and Y are indeed collinear. ∎

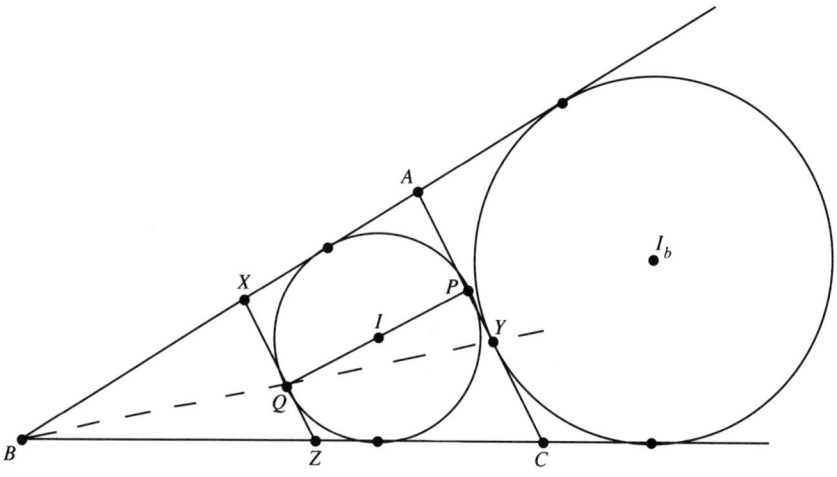

Figure 35

4. An Unlikely Concurrence

As usual, let I be the incenter of $\triangle ABC$ (Figure 36). Then AI is the bisector of angle A. If X and Y are the points of contact of the incircle on BC and AC, then

the lines AI, XY and the perpendicular from B to AI are concurrent.

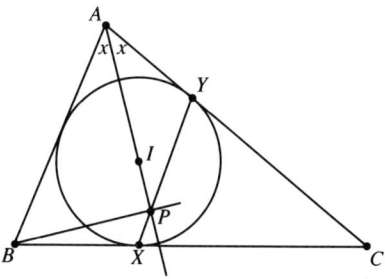

Figure 36

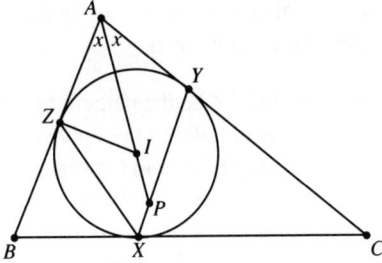

Figure 37

The easy proof involves only a few simple calculations. Letting P be the intersection of AI and XY, we shall show that BP is perpendicular to AP.

Now, the equal tangents CX and CY make $\triangle CXY$ isosceles, and each of the base angles at X and Y is $\frac{1}{2}(180° - C) = 90° - C/2$. If Z is the third point of contact (Figure 37), then similarly $\angle BXZ = 90° - B/2$, and we have

$$\angle ZXY = 180° - (90° - B/2) - (90° - C/2) = B/2 + C/2.$$

Since $\angle ZAP = A/2$, then in quadrilateral $ZAPX$ the opposite angles at A and X have the sum

$$\angle ZAP + \angle ZXP = A/2 + B/2 + C/2 = \text{a right angle.}$$

Consequently, the other pair of opposite angles, at Z and P, must add up to 270 degrees. Deducting 90 degrees for the right angle AZI, we have, in quadrilateral $ZIPX$, that the opposite angles at Z and P add up to 180 degrees, and we conclude that the quadrilateral is cyclic. That is to say, the circle around $\triangle ZIX$ also goes through the point P.

However, the right angles at Z and X (Figure 38) clearly show that $ZIXB$ is cyclic, and so the circumcircle of $\triangle ZIX$ goes through B, too.

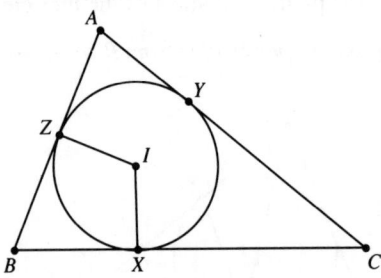

Figure 38

And because BZI is a right angle, BI is a diameter. In this case, BI subtends a right angle at the point P on the circumference, and the proof is complete. ∎

ON TRIANGLES

Exercise Set 3

3.1 In $\triangle ABC$, let the bisector of angle A meet BC at U. Prove that the perpendicular bisector of AU, the perpendicular to BC at U, and the circumdiameter through A are concurrent (Figure 39).

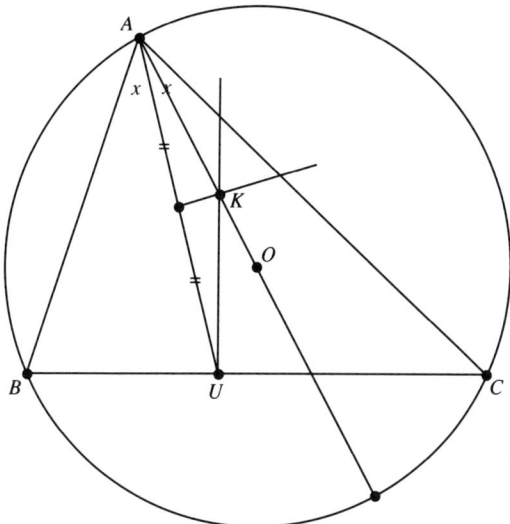

Figure 39

3.2 Let AD be the altitude from A in $\triangle ABC$ and let X and Y be the midpoints of the other two altitudes. Prove that the circumcircle of $\triangle DXY$ passes through the orthocenter H and the midpoint A' of BC. Also, prove that triangles DXY and ABC are similar (Figure 40).

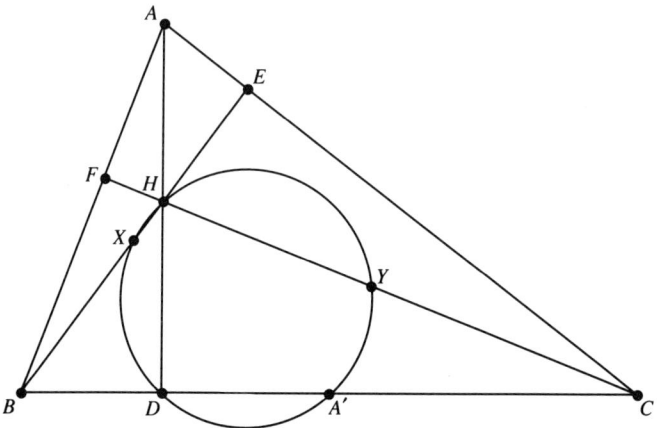

Figure 40

EXERCISES

34. In $\triangle ABC$ the bisector of angle A meets BC at P. Prove that the perpendiculars from B and C to the bisector AP are proportional to the perpendiculars through A to BP and CP respectively.

Figure 29

35. CAD being a straight line and D in $\triangle ABC$ such that DE is the bisector of the other two sides. Prove that the circle on BA as diameter passes through the intersection B and the perpendiculars B' at D. Also prove that angles DEF and BDC are equal.

Figure 30

CHAPTER FOUR

On Quadrilaterals

1. General Quadrilaterals

When you consider all the irregular shapes a quadrilateral can take, it really is quite remarkable that the midpoints of the sides always generate a figure with as much *order* as a parallelogram. Since this "Varignon" parallelogram of midpoints is discussed in many places, let us not pursue it here. Another instance of regularity emerging from even the most eccentric quadrilateral is the following result:

The four angle-bisectors of a quadrilateral always determine a cyclic quadrilateral.

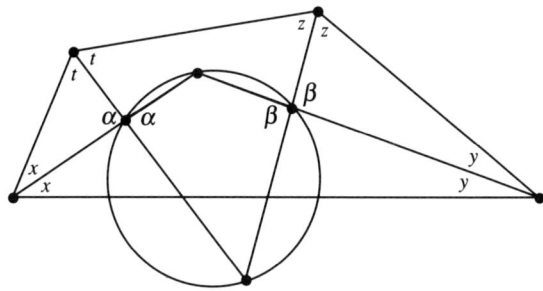

Figure 41

We compute the sum of a pair of opposite angles and find that it is 180°:
$$\alpha = 180° - (x + t) \quad \text{and} \quad \beta = 180° - (y + z),$$
hence
$$\alpha + \beta = 360° - (x + y + z + t)$$
$$= 360° - 180° = 180°. \blacksquare$$

2. Cyclic Quadrilaterals

Admittedly, confining our attention to *cyclic* quadrilaterals will greatly reduce the generality of our discoveries; however, quite a broad range of shapes will be left to us, and the specialization does give rise to many striking features. Let's begin with a remarkable property that is shared by all cyclic quadrilaterals.

(a) A REMARKABLE RESULT. *The four lines, each drawn from the midpoint of a side of a cyclic quadrilateral perpendicular to the opposite side, are concurrent* (Figure 42).

One promising way to approach the problem of concurrence is to try to come up with an independent specification of the proposed point of concurrence and show that it lies on each of the lines in question; for example, recall that the orthocenter of a triangle can be obtained independently of the altitudes by joining the circumcenter O to the centroid G and extending it twice the distance OG to a point H. Of course, this specification of the orthocenter was almost certainly discovered long after its existence had been established. Admittedly this technique might borrow a later discovery to establish an earlier one, but so long as this does not introduce a logical circularity, everything is all right (and besides that, it makes you look positively brilliant).

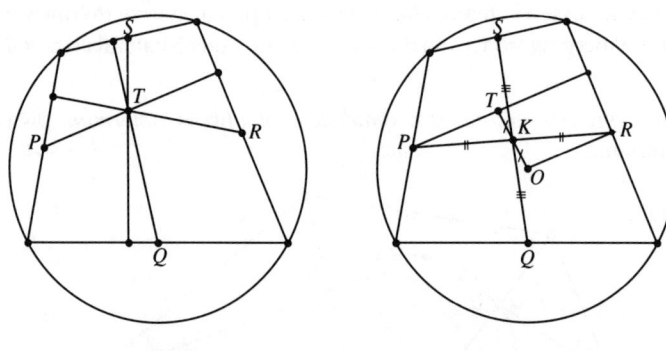

Figure 42   Figure 43

In the case at hand, then, let the lines PR and QS, which join the opposite pairs of midpoints, intersect at K and let the segment OK, from the center O, be extended its own length to a point T (Figure 43). We shall show that

the lines through T, from the midpoints P, Q, R, S, are each perpendicular to the opposite side.

We need only consider a typical case like PT.

As we noted above, the midpoints P, Q, R, S are the vertices of a parallelogram; hence K is the intersection of the diagonals and, as such, it bisects each of them. Thus the triangles PKT and KOR are congruent (two sides and the contained angle), and we have

$$\angle KPT = \angle KRO,$$

making PT parallel to OR. But because R is the midpoint of a chord, OR is perpendicular to the side; so then is PT. ∎

The points K and T are known as the *centroid* and *anticenter* of the given quadrilateral. (It is an easy exercise to show that K is the center of gravity of a

ON QUADRILATERALS 37

system of equal masses suspended at the vertices.) This result, then, can be stated in the form

a line joining the midpoint of a side of a cyclic quadrilateral to its anticenter is perpendicular to the opposite side.

(b) We have seen how to locate the anticenter of a cyclic quadrilateral $ABCD$ as the reflection of the circumcenter O in the center K of its Varignon parallelogram. The position of the anticenter in the particular case of a cyclic quadrilateral with perpendicular diagonals is the discovery of the Hindu mathematician Brahmagupta around the year 628 A.D. As you might have guessed, it is simply the point of intersection of the diagonals. In this light, the theorem might not seem to be a very remarkable result, but when considered on its own in the following form, it is a most striking proposition, indeed.

BRAHMAGUPTA'S THEOREM *In a cyclic quadrilateral having perpendicular diagonals, the perpendicular to a side from the point of intersection of the diagonals always bisects the opposite side.*

Rather than show formally that the intersection of the diagonals is actually the anticenter, thus involving the center and the centroid, it is much easier just to establish the result directly.

Let the acute angles in right triangle AET be x and y (Figure 44). Since the diagonals are perpendicular, $\angle DTE = x$, and in right triangle DET the angle at D is y. The pairs of vertically opposite angles at T then give the further angles x and y shown there. But the chord CD subtends equal angles x at A and B on the circumference, and chord AB similarly gives equal angles y at C and D. Thus the triangles TQC and TQB are both isosceles, and the arms QC and QB are equal, since each is equal to the common arm QT. ∎

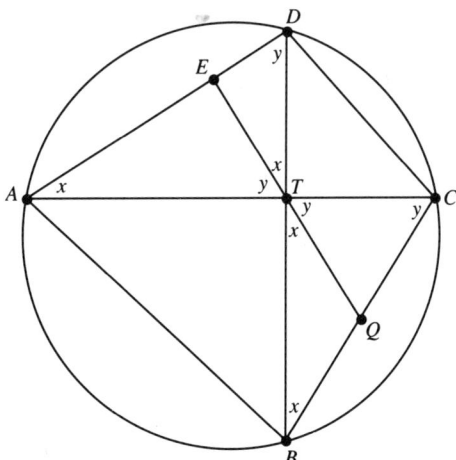

Figure 44

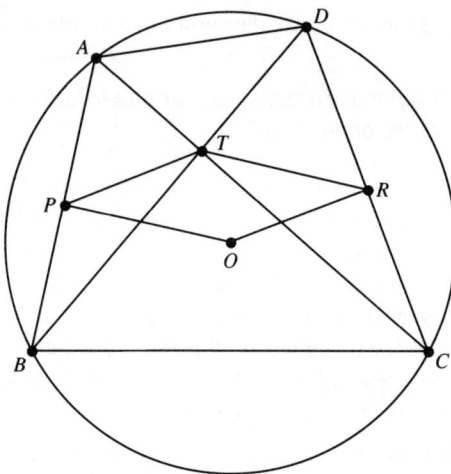

Figure 45

(c) A very nice application of Brahmagupta's theorem is the following property:

In a cyclic quadrilateral with perpendicular diagonals, the distance from the circumcenter to a side is just half the opposite side; i.e. in Figure 45,

$$OP = \frac{1}{2} \cdot CD.$$

The proof is easy and pretty.

The perpendicular from O clearly meets AB in its midpoint P. Now, by Brahmagupta's theorem, the line through the midpoint R of CD and the anticenter T of $ABCD$ is perpendicular to AB, and hence RT is parallel to OP. Similarly, $PT \parallel OR$ and $PORT$ is a parallelogram. Thus opposite sides OP and RT are equal. But R, being the midpoint of the hypotenuse of right triangle DTC, and thus its circumcenter, is equidistant from the three vertices, and we have

$$OP = TR = \frac{1}{2} \cdot CD. \blacksquare$$

This result solves the following problem most incisively.

(d) Suppose the diagonals of cyclic quadrilateral $ABCD$ are perpendicular and that E is the point on the circumcircle diametrically opposite vertex D. Prove that

$$AE = BC \qquad \text{(Figure 46)}.$$

Clearly chords at the same distance from the center of a circle are equal. From the previous result, we know that the distance from O to BC is $\frac{1}{2}AD$ (Figure 46). And it is easy to see that the perpendicular distance OM to AE is also $\frac{1}{2}AD$, since OM joins the midpoints of two sides of triangle ADE. $\blacksquare$

ON QUADRILATERALS

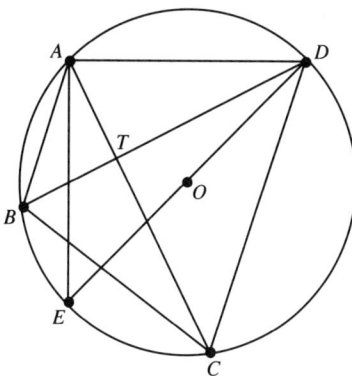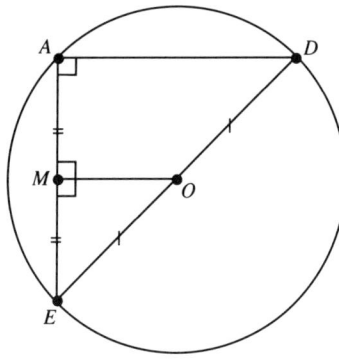

Figure 46

John Rigby pointed out that it would be simpler to proceed as follows. Since the length of a chord is characterized by the size of the angle it subtends at the circumference, we need only show that $\angle BAC = \angle ADE$. But this is virtually immediate, for

$\angle BAC = 90° - \angle ABD$ (from right triangle ABT)

$ = 90° - \angle AED$ (angles in the same segment of the circle)

$ = \angle ADE$ (from right triangle AED). ■

Let us conclude with the following beautiful result.

Let the diagonals of cyclic quadrilateral ABCD meet at M and let their midpoints be E and F (Figure 47). *Then the orthocenter of $\triangle EMF$ is the anticenter T of ABCD.*

Let P and Q be the midpoints of a pair of opposite sides, say AB and CD, as in Figure 47. First we establish a property noteworthy in its own right, namely that

the centroid K of ABCD is not only the midpoint of PQ (as we saw above) but is also the midpoint of the segment EF joining the midpoints of the diagonals.

This follows from the fact the $PFQE$ is a parallelogram: because of the midpoints, each of PE and FQ is parallel to BC and half as long as BC, making them equal and parallel. ■

Now, the anticenter is obtained by extending OK its own length to T. Thus a half-turn about K takes O to T and interchanges E and F. Accordingly, it takes OF to TE. But since F is the midpoint of chord BD, OF is perpendicular to BD. And since half-turns simply reverse the directions of lines, it follows that ET is also perpendicular to BD, making it the altitude from E in $\triangle EMF$. Similarly, FT is the altitude from F, and T is the orthocenter, as claimed. ■

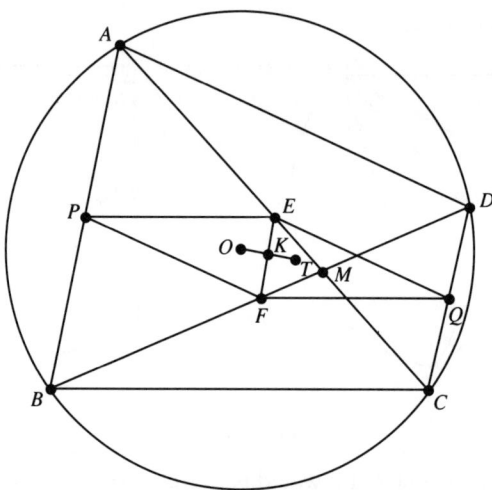

Figure 47

(Alternatively one can argue that triangles *OFK* and *KET* are congruent (two sides and the included angle), making alternate angles at *E* and *F* equal and *OF* parallel to *ET*. Since *OF* is perpendicular to *BD*, so is *ET*, making *ET* the altitude in △*EMF* from *E*. Similarly, *T* lies on the other altitudes and is therefore the orthocenter of △*EMF*.)

Comment

The centroid of a polygon is fundamentally defined as the center of mass of a system of equal masses suspended at the vertices. It is easy to deduce from this that the centroid *K* of our quadrilateral *ABCD* is located at both the midpoint of *PQ* and at the midpoint of *EF*, showing that these points are coincident:

(i) clearly a mass *m* at each of *A*, *B*, *C*, *D* is equivalent to a mass of 2*m* at *P* and a mass of 2*m* at *Q* (move the masses at *A* and *B* to **their** centroid, namely the midpoint *P* of *AB*, and the masses at *C* and *D* to **their** centroid, the midpoint *Q* of *CD*); thus the centroid *K* of the entire system is the midpoint of *PQ*.

(ii) similarly, by moving the masses at *A* and *C* to the midpoint *E* of *AC* and the masses at *B* and *D* to the midpoint *F* of *BD*, we have an equivalent sytem in equal masses of 2*m* at each of *E* and *F*, implying that *K* is also the midpoint of *EF*. ■

Note that the cyclic nature of *ABCD* is not invoked in this argument and therefore the result holds generally, even for self-intersecting quadrilaterals.

ON QUADRILATERALS

Exercise Set 4

4.1 Suppose opposite sides AD and BC of a cyclic quadrilateral meet at a point E (Figure 48). If Q and S are the midpoints of BC and AD, prove that the perpendicular from E to QS goes through the anticenter T.

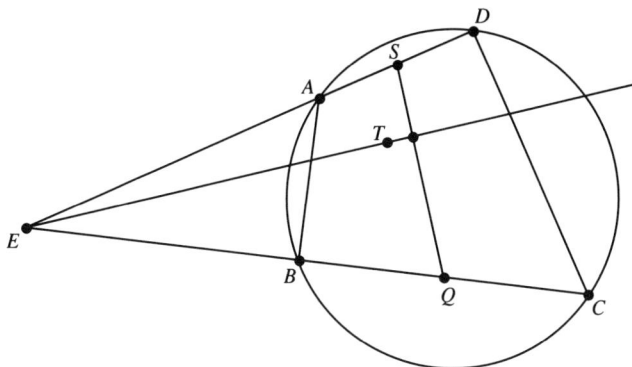

Figure 48

4.2 If the center O of a cyclic quadrilateral is reflected in a pair of opposite sides, prove that the line joining the images O_1 and O_2 always passes through the anticenter T (Figure 49).

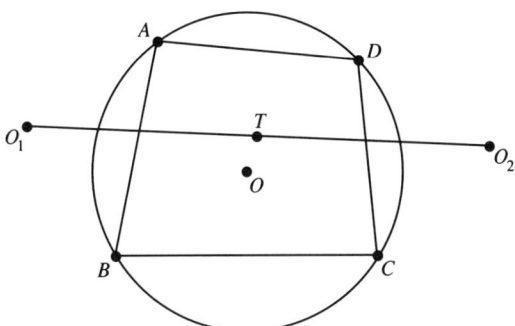

Figure 49

4.3 If the bisectors of one pair of opposite angles in a quadrilateral meet on a diagonal, prove that the bisectors of the other pair of opposite angles meet on the other diagonal.

CHAPTER FIVE

A Property of Triangles

1. The Property

Let P be a point on the circumcircle of triangle ABC. Reflect P in each side of the triangle to yield images P_1, P_2, and P_3 (Figure 50). Then, as strongly suggested by a figure,

the three images are collinear, and even more remarkably, *the line through them also passes through the orthocenter H of the triangle,* no matter where P might be chosen on the circumcircle.

In our proof of this result, we shall make use of the so-called "Simson line," and so let us digress briefly to establish this famous line.

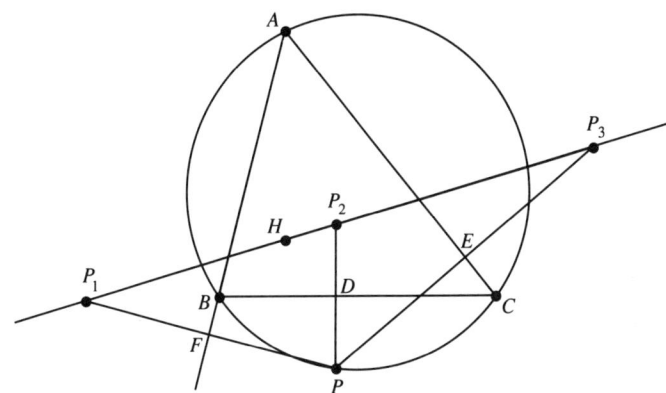

Figure 50

2. The Simson Line

In the process of reflecting a point P in the sides of $\triangle ABC$, a perpendicular is drawn from P to each of the sides, yielding the feet D, E, and F. It is a celebrated discovery that

these three feet are collinear if and only if P lies on the circumcircle of $\triangle ABC$, and the line they determine is called the *Simson line* of the point P.

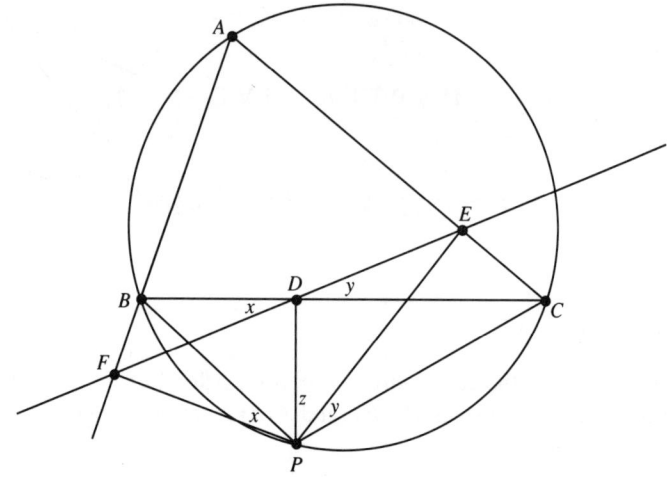

Figure 51

In Figure 51, join FD and DE, let the angles at D be called x and y, and let $\angle BPE = z$. We want to show that FD and DE are parts of the same straight line by showing that $x = y$. Since quadrilateral $BFPD$ is cyclic (the angles at F and D are right angles), $\angle FPB$ on chord FB is also equal to x. Similarly, in cyclic quadrilateral $DPCE$ (PC subtends right angles at D and E), $\angle CPE$ is equal to y. Now, clearly $ABPC$ is cyclic, making the opposite angles A and BPC supplementary:

$$A + z + y = 180°.$$

But the right angles at F and E make $AFPE$ cyclic and so we also have

$$A + x + z = 180°;$$

hence

$$x = y,$$

and FDE is the Simson line of P. ∎

The converse result follows simply by reversing this argument.

3. The Proof of the Property (John Rigby)

The collinearity of the images P_1, P_2, and P_3 follows immediately from the collinearity of the feet F, D, and E on the Simson line—clearly the straight line image of FDE under the dilatation $P(2)$ contains each of P_1, P_2, and P_3. The real problem is to show that the orthocenter H lies on $P_1P_2P_3$. However, since the dilatation $P(2)$ also shows that $P_1P_2P_3$ is parallel to the Simson line, we need show only that P_3H (or either of P_1H or P_2H) is parallel to the Simson line DE.

If altitude BH is extended to cross AC at K and the circle at J, then, as we have seen in Chapter 2, Section 2, $HK = KJ$. Let PJ cross AC at L and consider the

A PROPERTY OF TRIANGLES 45

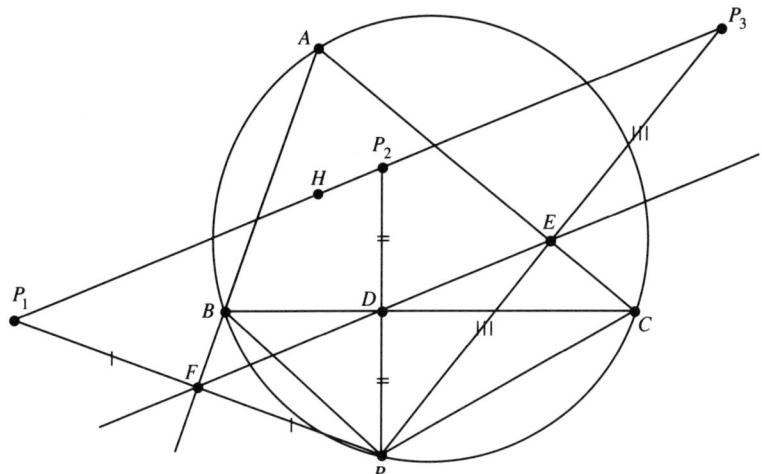

Figure 52

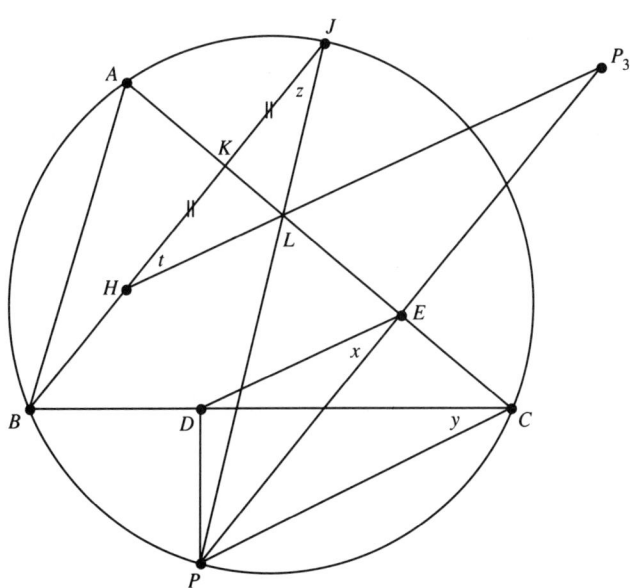

Figure 53

reflection of *PLJ* in the side *AC*. Clearly *P* goes to P_3, *J* to *H*, and *L* stays put. Thus *L* lies on the image HP_3, and the direction of HP_3 is given by the direction of *HL*.

Now, the altitude *BK* and the line PEP_3 are parallel (both are perpendicular to *AC*). Therefore, if we can show that *HL* is inclined to the one (*HK*) at the same angle that the Simson line *DE* is inclined to the other (*PE*), it will follow that *HL* and *DE* lie in the same direction. In Figure 53, then, we wish to show that the angles *x* and *t* at *E* and *H* are equal.

Observing that *PDEC* is cyclic (*PC* subtends right angles at *D* and *E*), we have $x = y$ on chord *DP*, and, in the circle shown in the figure, $y = z$ on chord *BP*. Thus $x = z$, and since *KL* is clearly the perpendicular bisector of *HJ*, we have $LH = LJ$, and $z = t$ in isosceles triangle *HLJ*. Hence $x = t$, and the proof is complete. ∎

4. A Corollary

Referring to Figure 52 (in the previous section), the Simson line *FD* joins the midpoints of two sides of $\triangle P_1P_2P$, and therefore *FD* bisects every segment from *P* to the side P_1P_2, in particular *PH*. Thus

the Simson line of a point P bisects the segment PH joining P to the orthocenter H of the given triangle ABC.

But we can take this even farther.

A LITTLE BONUS. Three of the special points on the nine-point circle of $\triangle ABC$ are the *Euler points* *X*, *Y*, and *Z*, which bisect the segments *AH*, *BH*, and *CH* (Figure 54). Consequently, the dilatation $H(\frac{1}{2})$ sends *A*, *B*, and *C*, into *X*, *Y*, and *Z*, respectively, and therefore takes the circumcircle to the nine-point circle. Now if *P* is any point on the circumcircle, we have just seen that the Simson line of *P* bisects *HP*. But clearly $H(\frac{1}{2})$ sends *P* to the midpoint of *HP*. Thus we can extend the above property of the Simson line to the more comprehensive result:

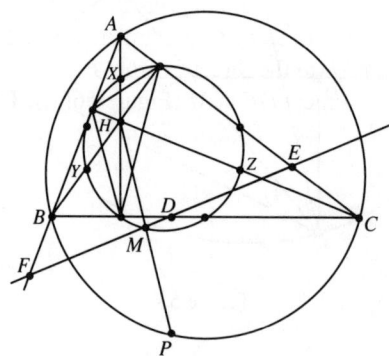

Figure 54

for a point P on the circumcircle, the midpoint of HP lies on both the Simson line of P and the nine-point circle.

5. A Property of Parabolas

Let us conclude with an engaging property of parabolas. Since no two tangents to a parabola have the same direction, *any* three tangents determine a triangle, and every such triangle enjoys the following property:

the circumcircle of the triangle formed by any three tangents to a parabola always goes through its focus.

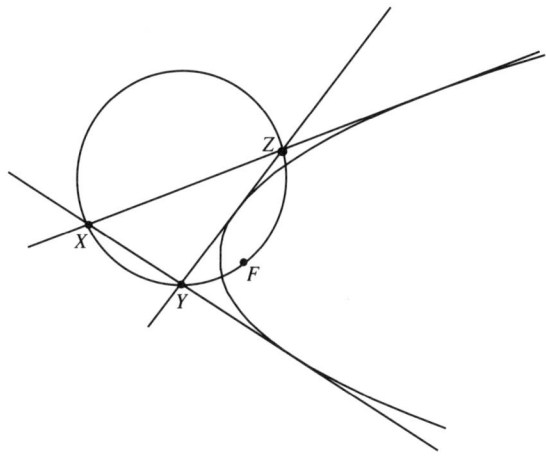

Figure 55

As a prelude to the easy proof of this result, let us review a few simple properties of parabolas.

(a) Recall that a point P on a parabola has the same perpendicular distance PD to the directrix as to the focus F. In particular, the vertex V of the parabola bisects the perpendicular FE to the directrix from F.

Now let the tangent at V meet DF at M (Figure 56(a)). Then, VM issues from the midpoint V of one side of $\triangle DEF$, is parallel to a second side (DE), and therefore bisects the third side, giving $DM = MF$. Consequently the triangles DPM and MPF have respectively equal sides and hence are congruent, yielding

(i) PM bisects the angle DPF, and
(ii) $\angle FMP = \angle DMP$, making FM perpendicular to PM.

(b) The reflector property states that

a ray from the focus to a point P on the parabola is reflected in the direction of the axis.

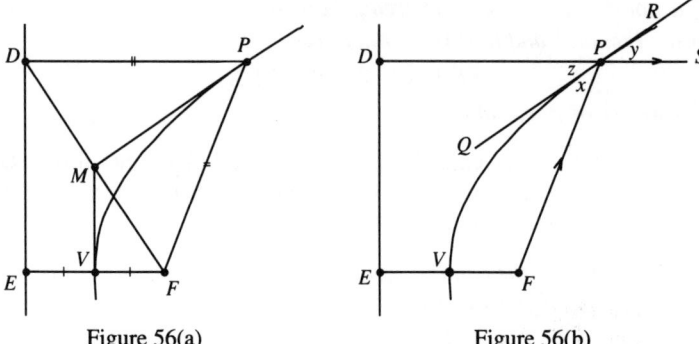

Figure 56(a)　　　　　Figure 56(b)

Let QPR be the tangent at P. Since rays are reflected at P exactly as they would be reflected by the *tangent* at P, the reflector property says that the angles x and y at P in Figure 56(b) are equal.

Since the vertically opposite angles y and z are equal, we have $x = z$; hence the bisector of angle DPF is the tangent at P. Thus in Figure 56(a), PM is the tangent at P. Having already observed that FM is perpendicular to MP, we obtain the important result that

the foot M of the perpendicular to a tangent to a parabola from the focus F always lies on the tangent at the vertex V.

(c) By the result just proved, the feet of the perpendiculars S, U, W from the focus F to the sides of triangle XYZ formed by three tangents lie on the tangent TV at the vertex (Figure 57). Thus TV is *the Simson line* of the point F with respect to the triangle XYZ, and since only points on the circumcircle of a triangle have Simson lines, it follows that the circumcircle of a triangle determined by three tangents to a parabola passes through its focus. ∎

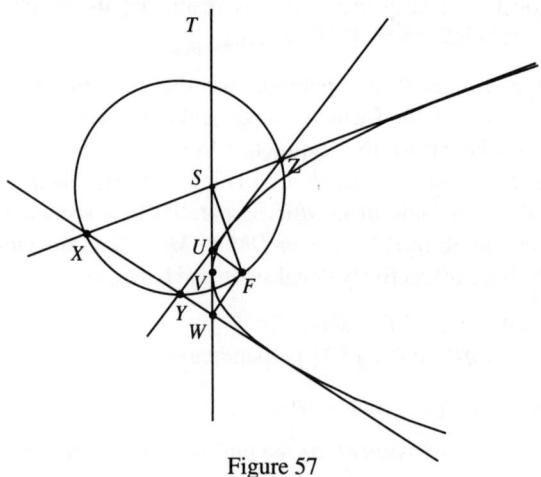

Figure 57

CHAPTER SIX

The Fuhrmann Circle

Suppose the circumcircle is drawn around a given triangle ABC and that X', Y', and Z' are the midpoints of the arcs cut off by the sides (Figure 58). Now suppose that each of these midpoints is reflected in the nearest side of the triangle. Then the triangle of images XYZ and its circumcircle are called the *Fuhrmann triangle* and the *Fuhrmann circle* of $\triangle ABC$, after the 19th-century German geometer Wilhelm Fuhrmann (1833–1904).

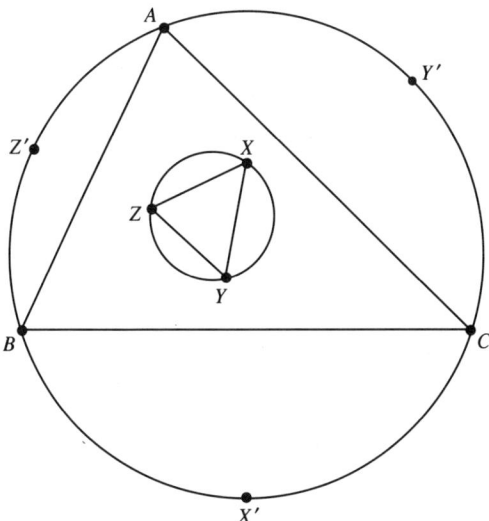

Figure 58

In earlier essays we have occasionally referred to the famous nine-point circle of a triangle. The Fuhrmann circle is of the same status—it goes through at least 8 noteworthy points of the triangle. For example, the Fuhrmann circle always goes through the orthocenter H and also through the Nagel point M. We have already seen that H and M are both related to the centers of one of the special circles of a triangle:

- the center of the nine-point circle is the midpoint of the segment HO joining the orthocenter to the circumcenter, and

- the center S of the Spieker circle (i.e. the incircle of the medial triangle) is the midpoint of the segment IM joining the incenter to the Nagel point.

Now, in the Fuhrmann circle the points H and M are brought together as equal partners, for

the center F of the Fuhrmann circle is the midpoint of HM itself, implying that HM is always a diameter of the Fuhrmann circle.

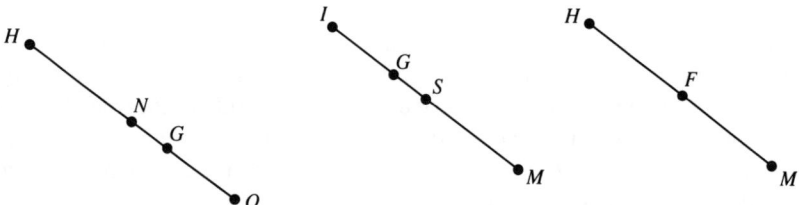

Figure 59

PROOF (John Rigby). (a) Since X' is the midpoint of the arc BC, its image X lies on the perpendicular bisector of BC, i.e. on the diameter $X'OL$ (Figure 60(a)). The extended altitude $AHDD'$ is also perpendicular to BC and therefore is parallel to LOX'. Consequently, reflection in the perpendicular diameter KOJ, takes L to X' and A to D' (hence AL to $D'X'$), and also takes angle α to β.

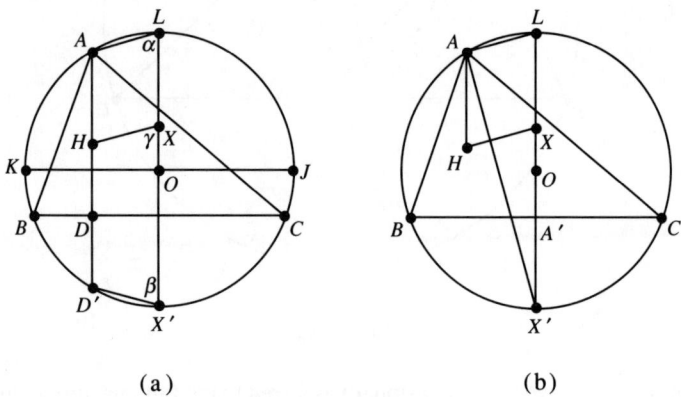

Figure 60

Recalling from Chapter 2, on the orthocenter, that $HD = DD'$, a further reflection in BC takes $D'X'$ to HX and angle β to γ. Thus

$$AL = D'X' = HX \quad \text{and} \quad \alpha = \beta = \gamma,$$

so AL is equal and parallel to HX.

Since LX' is a diameter, $\angle LAX'$ is a right angle (Figure 60(b)), and since HX is parallel to AL, it follows that *HX is perpendicular to AX'*.

(b) In Chapter 1, on cleavers and splitters (Section 1c), we encountered the following property of the Nagel point:

the incenter of a triangle is the Nagel point of its medial triangle.

Now, a triangle PQR, having sides twice as long as those of the given triangle ABC, is obtained by drawing lines through the vertices A, B, and C parallel to the sides opposite these vertices (Figure 61). This creates parallelograms $RBCA$ and $ABCQ$ in which $RA = BC = AQ$, making A the midpoint of RQ. Similarly, B and C are the midpoints of PR and PQ, and $\triangle ABC$ is the medial triangle of $\triangle PQR$. Accordingly, the Nagel point M of $\triangle ABC$ is the incenter of $\triangle PQR$, and therefore MP is the bisector of $\angle P$.

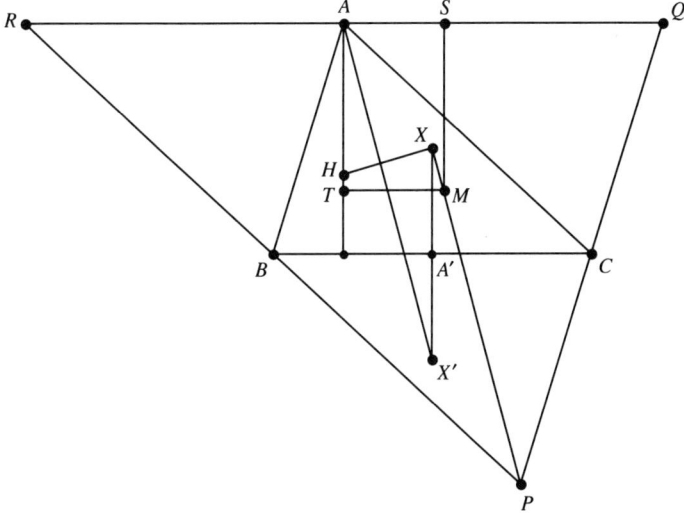

Figure 61

Now, in parallelogram $ABPC$, the diagonals AP and BC bisect each other at the midpoint A' of BC. As a result, a half-turn about A' takes AX' to XP, and therefore AX' and XP are parallel. But, since X' bisects the arc BC of the circumcircle of $\triangle ABC$, AX' is the bisector of $\angle A$ in $\triangle ABC$. Thus AX' is an angle bisector in the medial triangle of PQR and, by a result established in Chapter 1 on cleavers and splitters, AX' is therefore parallel to the corresponding angle bisector MP of $\triangle PQR$. Thus AX' is parallel to both XP and MP, and we conclude that X, M and P are collinear.

(c) Since *HX* is perpendicular to *AX'*, which is itself parallel to *XMP*, it follows that *HX* is perpendicular to *XM*, that is,

$$\angle HXM = \text{a right angle},$$

and the circle on *HM* as diameter contains *X*. Similarly, this circle also passes through *Y* and *Z* and is therefore the Fuhrmann circle, as claimed. ∎

A COROLLARY. Referring to Figure 61, the length of the perpendicular *MS* from the incenter *M* of $\triangle PQR$ to the side *QR* is clearly the inradius of $\triangle PQR$. The perpendicular *MT* to the extended altitude *AH* completes a rectangle *ATMS* in which *AT* equals the inradius *MS*. Since *HM* subtends a right angle at *T*, the Fuhrmann circle also goes through *T*, implying that the Fuhrmann circle intersects an altitude at a point whose distance from the associated vertex is equal to the inradius of $\triangle PQR$ (the same is true of the other altitudes). But the sides of $\triangle PQR$ are twice as long as those of $\triangle ABC$; hence the inradius *MS* is twice the inradius *r* of $\triangle ABC$.

Thus three additional points of note on the Fuhrmann circle are the points T, U, and V at distance 2r from A, B, C, along the altitudes from them (Figure 62).

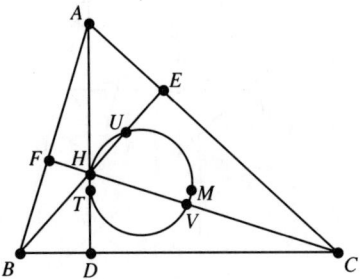

Figure 62

CHAPTER SEVEN

The Symmedian Point

1. It would be an understatement even to say that the symmedian point of a triangle is not very well known; in fact, most mathematicians in the present generation have never even heard of it. I prefer to think that this engaging subject is simply one of the casualties of the 20th-century explosion of learning. We don't have time to teach everything that has been discovered over the centuries, and this is one of the many worthy topics that never became part of our standard school or college curricula. I shouldn't like to think it lost out on the grounds that it wasn't very exciting, for the symmedian point is one of the crown jewels of modern geometry.

2. Isogonal Lines and Points

Two lines AS and AT through the vertex A of an angle are said to be *isogonal*, or *isogonal conjugates*, if they are equally inclined to the arms of $\angle A$, or equivalently to the bisector of $\angle A$ (Figure 63).

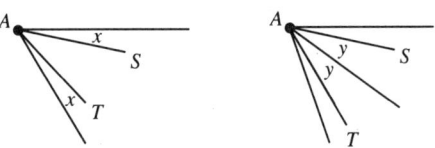

Figure 63

If a point P of $\triangle ABC$ is joined to each vertex, the line PA has an isogonal conjugate at A, PB has one at B and PC one at C (Figure 64). Now there is no immediately obvious reason why these three conjugates should be concurrent. We begin, then, with the basic property that this is always the case.

If lines through A, B, and C are concurrent at P, then the isogonal lines are concurrent at a point Q.

The points P and Q are also called *isogonal conjugates*.

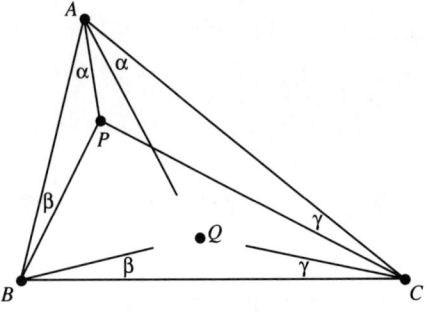

Figure 64

This property is an immediate consequence of the following even more fundamental result:

THEOREM 1. *The points P and Q lie on lines isogonal with respect to $\angle A$ if and only if the pair of distances from P to the sides of $\angle A$ are inversely proportional to the corresponding pair of distances from Q; that is, with labels as in Figure 65,*

$$\frac{x}{y} = \frac{s}{r}.$$

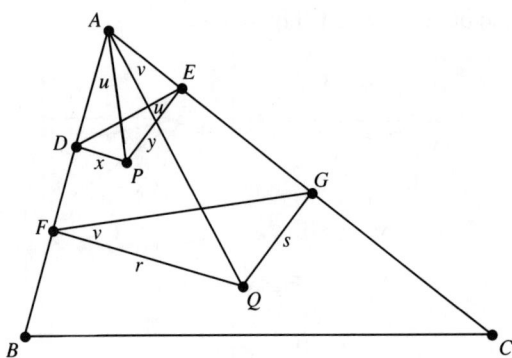

Figure 65

(a) *Sufficiency*: Suppose $x/y = s/r$.

In Figure 65, the right angles at D and E make $ADPE$ a cyclic quadrilateral, and we have

$$\angle DAP = \angle DEP = u \quad \text{on chord } DP, \quad \text{and} \quad \angle DPE = 180° - \angle A;$$

THE SYMMEDIAN POINT

similarly, the right angles at F and G make $AFQG$ cyclic, and

$$\angle QFG = \angle QAG = v \quad \text{on chord } QG, \quad \text{and} \quad \angle GQF = 180° - \angle A.$$

Thus, if $x/y = s/r$, the triangles DPE and FQG are similar, and we have equal angles u and v (opposite the corresponding sides x and s); hence AP and AQ are isogonal conjugates at A, as claimed.

(b) *Necessity*: Conversely, if AP and AQ are isogonal conjugates at A, then $u = v$, and triangles DEP and GFQ have equal corresponding angles and are therefore similar. The required proportion then follows immediately. ∎

Now we can proceed with the easy proof of the basic property.

Theorem 2. *In $\triangle ABC$, the isogonal conjugates of AP, BP, and CP are concurrent.*

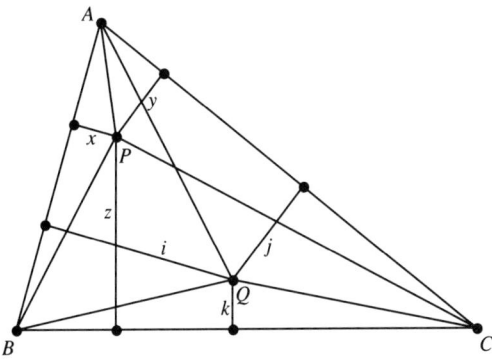

Figure 66

Let Q be the point of intersection of the conjugates of AP and BP; we will show that P and Q are also on conjugate lines from vertex C. Let the distances to the sides of $\triangle ABC$ from P and Q be $x, y, z; i, j, k$, as shown in Figure 66. Then, by our previous result, we have from vertex A that

$$\frac{x}{y} = \frac{j}{i},$$

and from B that

$$\frac{z}{x} = \frac{i}{k}.$$

Multiplying we get

$$\frac{z}{y} = \frac{j}{k},$$

which implies that PC and QC are isogonal lines at C. ∎

THREE BRIEF NOTES.

(i) Note that if Q is the isogonal conjugate of P, then P is the isogonal conjugate of Q; conjugate points occur in pairs.

(ii) Throughout our discussion, nothing is lost by thinking in terms of conjugate pairs (P, Q) that lie *inside* the parent triangle ABC, but we should note in passing that this is an unnecessary restriction. At each vertex V of $\triangle ABC$, the requirement of conjugacy can be stated in the following terms: measuring *from* the sides of the triangle,

the angle of inclination of PV to one side of the triangle at vertex V
$= -$(the angle of inclination of QV to the other side at V).

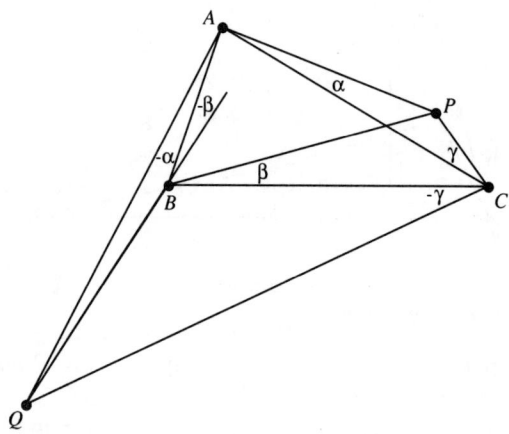

Figure 67

Exercise

Let P be a point of the circumcircle of $\triangle ABC$, prove that the lines AQ, BQ, and CQ to its conjugate Q are parallel, thus placing Q at infinity.

(iii) Observe also that the ratio of the distances to the sides of an angle A is the same for all points on a line through A (Figure 68):

THE SYMMEDIAN POINT

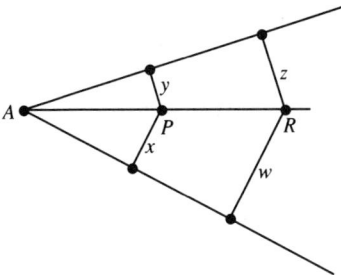

Figure 68

similar triangles give

$$\frac{x}{w} = \frac{AP}{AR} = \frac{y}{z} \implies \frac{x}{y} = \frac{w}{z}.$$

3. The Symmedians and the Symmedian Point K

The isogonal conjugates of the *medians* of a triangle are called *symmedians*, and their point of concurrency—i.e. the isogonal conjugate of the centroid G—is called the *the symmedian point K*. In Britain and France this point is also called the Lemoine point, and in Germany it is also called Grebe's point. You will find the early history and an extensive account of this point in two articles by John Mackay [1, 2].

Each line through the vertex A of $\triangle ABC$ determines its own ratio x/y of distances from any point on it to the sides AB and AC (Figure 69). It turns out that this ratio is a convenient way of identifying a symmedian.

THEOREM 3. *AP is a symmedian if and only if the distances to the sides AB and AC are proportional to the sides themselves*:

$$\frac{x}{y} = \frac{c}{b}.$$

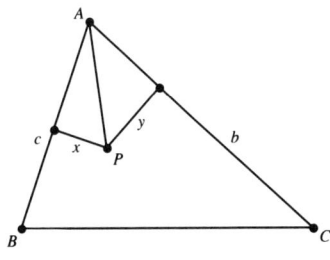

Figure 69

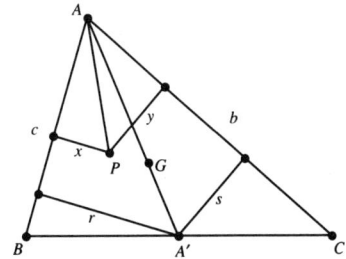

Figure 70

(a) *Suppose AP is a Symmedian*

In this case the median AA' and the line AP are a pair of isogonal lines at A, giving $x/y = s/r$ by Theorem 1 (Figure 70). But a median bisects the area of a triangle, and so r and s are altitudes in the triangles ABA' and $AA'C$ of equal area, and we have

$$\frac{1}{2}cr = \frac{1}{2}bs, \quad \text{giving} \quad \frac{s}{r} = \frac{c}{b}.$$

Therefore $x/y = s/r = c/b$, as claimed. ∎

(b) *Suppose $x/y = c/b$ for a line AP*

Referring to Figure 70, we have just seen that, for median AA',

$$\frac{s}{r} = \frac{c}{b}.$$

If $x/y = c/b$ for the line AP, then $x/y = s/r$, which implies, by Theorem 1, that AP and AA' are isogonal conjugates, making AP a symmedian.

A COROLLARY. It is well known that the bisector of an angle of a triangle divides the opposite side in the ratio of the sides about the angle. Well, a symmedian does it in the ratio of the *squares* of the sides:

THEOREM 4.

$$\frac{BP}{PC} = \frac{c^2}{b^2}.$$

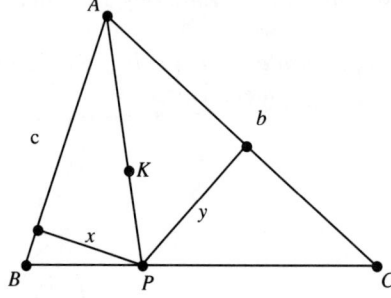

Figure 71

The proof is immediate. We have just seen, in part (a), that $x/y = c/b$ for symmedian AP, and therefore

$$\frac{BP}{PC} = \frac{\text{Area}(\triangle ABP)}{\text{Area}(\triangle APC)} = \frac{\frac{1}{2} \cdot cx}{\frac{1}{2} \cdot by} = \left(\frac{c}{b}\right)\left(\frac{x}{y}\right) = \frac{c^2}{b^2}. \quad \blacksquare$$

A Characterization of the Symmedian Point K

For the symmedian point K, the distances x, y, z to all three sides are proportional to the sides themselves:

(*) $\qquad x : y : z = a : b : c, \quad$ or $\quad \dfrac{x}{a} = \dfrac{y}{b} = \dfrac{z}{c}.$

Conversely, any point with this property must lie on all three symmedians—forcing it to coincide with the symmedian point K. Thus the proportion (*) characterizes the symmedian point uniquely.

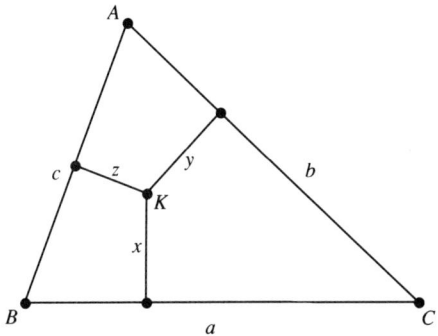

Figure 72

Now let's look at some pleasing applications of these basic theorems.

4. Applications and Further Developments

(i) The symmedian point K of a right angled triangle ABC is the midpoint of the altitude to the hypotenuse.

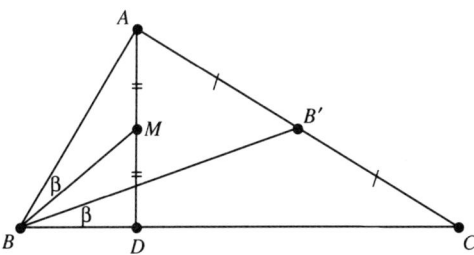

Figure 73

PROOF (John Rigby). Having equal corresponding angles, triangles DBA and ABC are similar (Figure 73). Thus triangle ABC with its median BB' is just an enlarged copy of triangle *DBA* with its corresponding median *BM* through their common vertex *B*. Accordingly, *BM* is inclined to the hypotenuse *BA* in *ABD* at the same angle β as median BB' is inclined to the hypotenuse *BC* in *ABC*. Hence *BM* is the symmedian of $\triangle ABC$ at *B*. Similarly, *M* lies on the other symmedians and is therefore the symmedian point of $\triangle ABC$. ∎

(ii) *The symmedians of a triangle bisect the sides of its orthic triangle.*

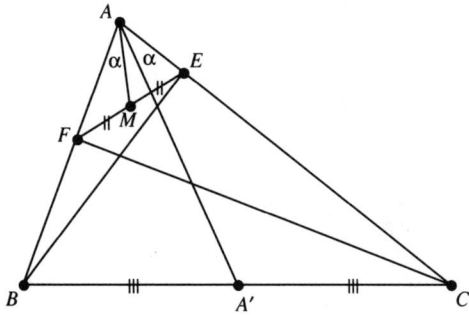

Figure 74

Recall from Chapter 2, Section 3a that the vertices of the orthic triangle *DEF* of $\triangle ABC$ are the feet of the altitudes, and that each side of the orthic triangle cuts off, at each vertex of $\triangle ABC$, a little triangle similar to *ABC*. In Figure 74, then, $\triangle AEF$ is similar to $\triangle ABC$. Because of this, the "similar triangle" approach that was just used in part (i) can also be applied here.

In these similar triangles it is $\angle AEF$ that is equal to angle *B* in $\triangle ABC$ (each is the supplement of $\angle FEC$), making *AF* and *AC* a pair of corresponding sides. Thus the medians *AM* and AA' from their common vertex *A* are inclined at the same angle α to their corresponding sides *AF* and *AC*, and hence *AM* is the symmedian of $\triangle ABC$ at *A*. ∎

(iii) Here is a very useful property of symmedians.

The tangents to the circumcircle of a triangle at two of its vertices meet on the symmedian from the third vertex.

In Figure 75, *AT* is the symmedian at *A*.

Again we can use the "similar triangle" approach of parts (i) and (ii).

Because of the theorem "the angle between a tangent and a chord is equal to the angle in the segment on the opposite side of the chord" and the equality of vertical angles, the angles β and γ of $\triangle ABC$ occur at the vertices as marked in Figure 75. If the circle $T(TB)$ meets *AB* extended at *X*, then $\triangle BXT$ is isosceles with base angles equal to γ.

THE SYMMEDIAN POINT

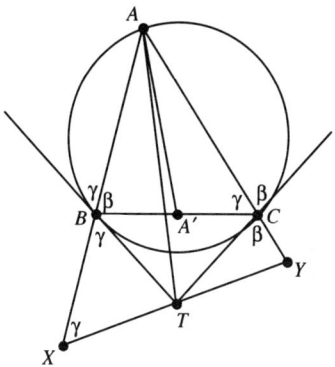

Figure 75

Now suppose XT meets AC extended at Y. Then, with angle A common, triangles ABC and AXY have two pairs of equal angles, making $\angle XYC = \beta$ and triangle TYC isosceles. Therefore

$$\begin{aligned} XT &= BT \quad \text{(in isosceles } \triangle XBT) \\ &= TC \quad \text{(equal tangents)} \\ &= TY \quad \text{(in isosceles } \triangle TYC), \end{aligned}$$

establishing AT as a median of $\triangle AXY$.

Thus, as in the previous cases, angles XAT and YAA' are the equal angles between corresponding medians and corresponding sides at the common vertex of similar triangles AXY and ABC, and the conclusion follows. ∎

(iv) The Gergonne Point

The *Gergonne point* of a triangle is the point of concurrence of the lines which join the vertices to the points of contact of the incircle on the opposite sides (Figure 76).

As in the case of the Nagel point, this concurrence is an immediate consequence of Ceva's theorem. Recall that Ceva's theorem asserts:

Three segments AD, BE, and CF from the vertices to the opposite sides of $\triangle ABC$ are concurrent if and only if the points D, E, and F divide the sides in ratios whose product is 1.

(For a proof of Ceva's theorem, see Chapter 12 On Cevians.) Since the two tangents to the incircle from a vertex are equal, we have immediately that

$$\frac{BD}{DC} \cdot \frac{CE}{EA} \cdot \frac{AF}{FB} = \frac{y}{z} \cdot \frac{z}{x} \cdot \frac{x}{y} = 1,$$

and the Gergonne point is established. ∎

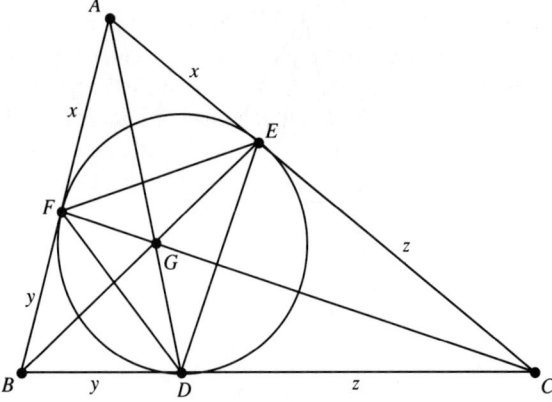

Figure 76

Let us call the triangle DEF, determined by the points of contact, the *Gergonne triangle* of $\triangle ABC$ (Figure 76). The incircle of $\triangle ABC$ is the circumcircle of $\triangle DEF$, so by the result of part (iii), the tangents at its vertices D, E, and F, give its symmedians DA, EB, and FC. Thus we have the delightful result that

the Gergonne point of a triangle is the symmedian point of its Gergonne triangle.

(v) A Real Gem

Suppose three lines are drawn through the Gergonne point G of $\triangle ABC$, one parallel to each side of its Gergonne triangle DEF. This determines a total of six points P, Q, R, S, T, U on the sides of $\triangle ABC$ (Figure 77). In 1843, C. Adams established the marvelous result that

these six points always lie on a circle, and moreover, this circle is concentric with the incircle.

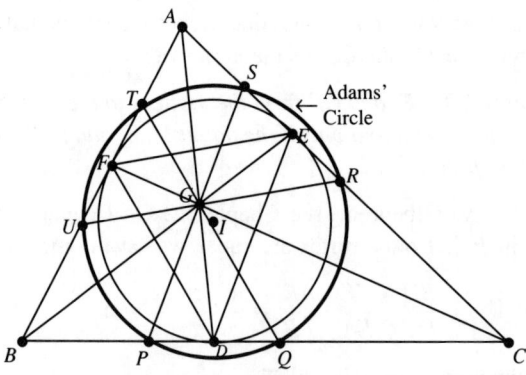

Figure 77

THE SYMMEDIAN POINT

PROOF (John Rigby). (a) We would like to show that each of the six points P, Q, R, S, T, U is at the same distance from the incenter I. Clearly the inradii ID, IE, and IF meet the sides of $\triangle ABC$ at right angles (Figure 78). Thus at D for example, the right triangles IDP and IDQ have a common leg $ID = r$. If we can show that PD and DQ are equal, then these right triangles are congruent and $IP = IQ$. More generally, if we can show that all six of the little segments PD, DQ, RE, ES, TF, and FU are equal, then six congruent right triangles result, and their six equal hypotenuses yield the desired conclusion. Therefore let us proceed to show that

$$PD = DQ = RE = ES = TF = FU.$$

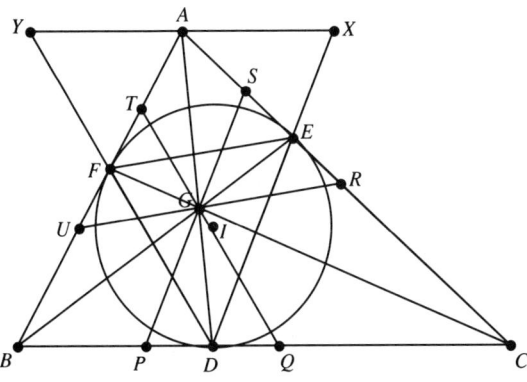

Figure 78

(b) The equal tangents CD and CE make $\triangle CDE$ isosceles, and the parallel lines DE and PS make triangles CDE and CPS similar. Thus $\triangle CPS$ is also isosceles, and the differences PD and SE in the lengths of the arms of these isosceles triangles are equal:

$$PD = SE.$$

Similarly $ER = FU$ and $FT = DQ$.

Thus, if we can show that $PD = DQ$, similar arguments will yield $RE = ES$ and $TF = FU$, and

$$PD = SE = ER = FU = FT = DQ$$

(Figure 78).

Let us conclude, then, by showing that $PD = DQ$.

(c) To this end, let extensions of DE and DF meet the line through A parallel to BC at X and Y (Figure 78). Triangles AXE and CDE have equal corresponding angles and therefore are similar; and since △CDE is isosceles, so is △AXE. Similarly, △AYF is similar to isosceles triangle BDF and, with the equal tangents AE and AF, we have altogether that

$$AX = AE = AF = AY,$$

revealing that DA is a median of △DXY (Figure 79).

Now the dilatation $D(\frac{DG}{DA})$ sends the equal segments YA and AX to equal images MG and GN which lie along a segment of MGN parallel to YX, and therefore also parallel to PQ (figure 79). Thus, in parallelograms MGQD and GNDP, we have

$$MG = DQ \quad \text{and} \quad GN = PD,$$

and since $MG = GN$, then $PD = DQ$, completing the proof.

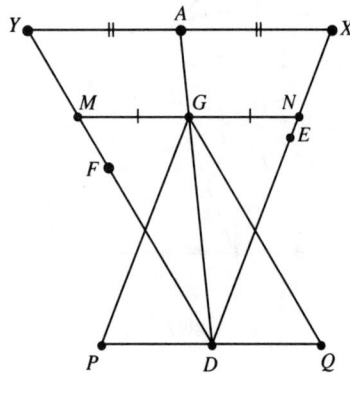

Figure 79

(vi) Let's pick up one final property before going on to further applications.

Suppose X is a point on either of two isogonal lines AL and AM through A (Figure 80). Then the line YZ joining the feet of the perpendiculars from X to AB and AC is perpendicular to the other line of the pair.

The proof is quick and easy. Because the lines are isogonal, the angles α at A are equal. But clearly AYXZ is cyclic, so $\angle YZX = \angle YAX = \alpha$. It remains only to observe that because the one pair of arms AZ and XZ of the equal angles YZX and MAZ are perpendicular, the other pair of arms, namely YZ and AM, lying at equal directed angles from AZ and XZ, respectively, are also perpendicular.

THE SYMMEDIAN POINT

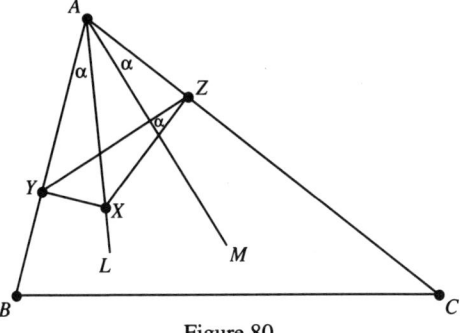

Figure 80

(vii) Here's a pretty result.

The line from the midpoint of a side of a triangle to the midpoint of the altitude to that side goes through the symmedian point;

in Figure 81, $A'M$ goes through K.

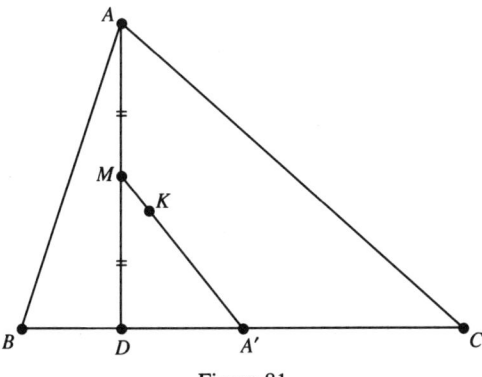

Figure 81

PROOF (John Rigby). Dilatations often provide us with a nice way of relating similar triangles, and we shall repeatedly take advantage of this more sophisticated point of view here.

Let the tangents to the circumcircle of $\triangle ABC$ at the vertices determine $\triangle XYZ$ (Figure 82). By part (iii) of this chapter, AX, BY, and CZ are the symmedians of $\triangle ABC$, and therefore each of them goes through K. Let lines through K, parallel to the sides of $\triangle XYZ$, give the six points P, Q, R, S, T, and U on the sides of $\triangle ABC$, as shown (shades of Adams' Circle). As we shall see, these six points lie on a circle with center K (a circle we shall recognize in a later chapter to be the Second Lemoine Circle of $\triangle ABC$).

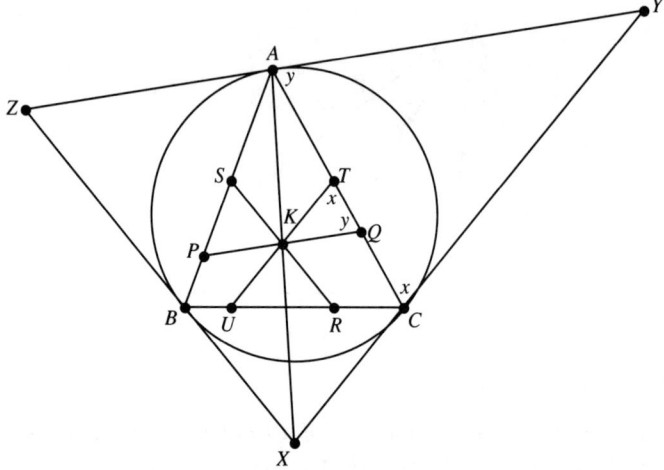

Figure 82

In $\triangle ABX$, SK is parallel to BX, and therefore S and K divide AB and AX in the same ratio; similarly, in $\triangle ACX$, TK is parallel to CX, and so T divides AC in the same ratio as K divides AX. Thus the dilatation $A(\frac{AK}{AX})$ takes XB to KS and XC to KT. But XB and XC are equal tangents from X, and therefore their images must also be equal:

$$KS = KT.$$

Similarly, two other pairs of equal segments emanating from K are

$$KQ = KR \quad \text{and} \quad KU = KP.$$

Now consider KT from one of these pairs and KQ from another. One set of parallel lines gives equal alternate angles x at T and C, and another pair of parallels gives equal alternate angles y at Q and A. But the equal tangents YA and YC make $\triangle AYC$ isosceles and we have $x = y$. Hence $\triangle TKQ$ is isosceles and $KT = KQ$. A similar argument gives $KR = KU$, and it follows that all six of the segments are equal:

$$KS = KT = KQ = KR = KU = KP,$$

thus establishing a circle through P, Q, R, S, T, U with center K.

Since SR and TU are diameters of this circle, $STRU$ is a rectangle, and with sides TR and SU perpendicular to UR (i.e. to BC), they are therefore parallel to the altitude AD in $\triangle ABC$ (Figure 83).

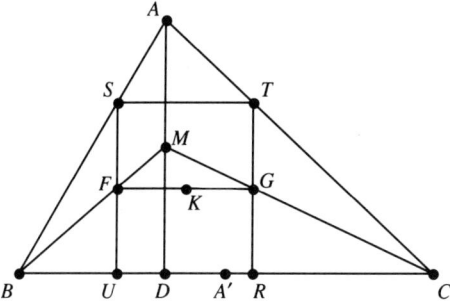

Figure 83

Now let M be the midpoint of AD, and F and G the midpoints of the sides SU and TR. At this point we do not know that M, F and B are collinear. However, the midline FG bisects the rectangle, is parallel to UR (i.e. to BC), passes through the center K of the rectangle, and is bisected there. Since SU is parallel to AD, the dilatation $B\left(\frac{BS}{BA}\right)$ takes AD to SU and accordingly takes its midpoint M to the midpoint F of SU. It follows, then, that B, F, and M lie on a straight line. Similarly, M, G, and C are collinear.

Finally, since FG is parallel to BC, the dilatation $M\left(\frac{MF}{MB}\right)$ takes BC to FG, and in the process takes the midpoint A' of BC into the midpoint K of FG. Hence M, K, and A' are collinear. ∎

(viii) The pedal circle

The *pedal triangle* of a point P, with respect to $\triangle ABC$, is the triangle determined by the feet of the perpendiculars from P to the three sides of $\triangle ABC$. Thus every point P in the plane of the triangle has a pedal triangle, and the *orthic triangle* of $\triangle ABC$ is simply the pedal triangle of its orthocenter H. Now let's consider a remarkable result.

The six vertices of the pedal triangles of a pair of isogonal points (X, P) lie on a circle whose center is the midpoint M of the segment XP (Figure 84).

That is to say, conjugate points have pedal triangles which have a common circumcircle; this circle is called the *pedal circle* of (X, P).

We shall show that there is a circle through the four feet R, Z, Y and Q (Figure 84), and that its center is the midpoint M of XP. Similarly, the feet Z, R, U, and V also lie on a circle with center M, obviously the same circle $M(MZ)$ since Z and R are common.

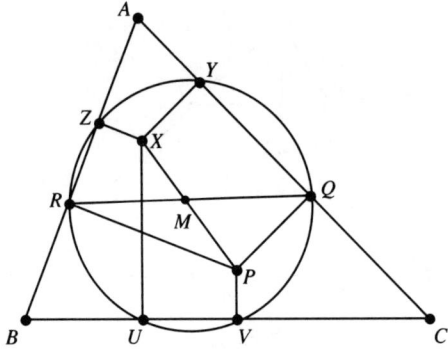

Figure 84

Our proof is based on the following theorem:

(a) Let *PAD* and *PBC* be two secants to a circle from a point *P* outside the circle, then

the product of the lengths of the whole secant and the part outside the circle is the same for both secants (Figure 85):

$$PD \cdot PA = PC \cdot PB$$

(a result deduced from similar triangles *PCD* and *PAB*).

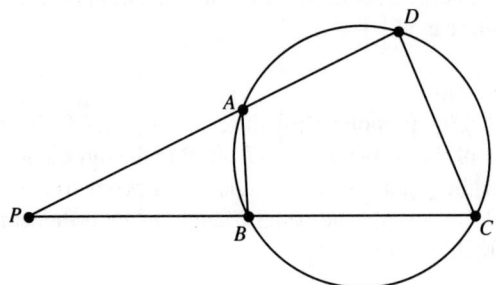

Figure 85

(b) Conversely, let *A* lie on *PD*, *B* on *PC* such that $PD \cdot PA = PC \cdot PB$; then *ABCD* is a cyclic quadrilateral.

Since *X* and *P* are on isogonal lines from *A*, the "perpendicular" property established in part (vi) above implies that the line *RQ* which joins the feet of the perpendiculars from *P* to *AB* and *AC* is perpendicular to *AX*, giving right angles at *S* (Figure 86).

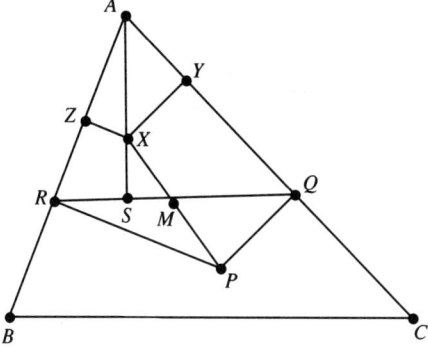

Figure 86

With right angles at S and Z, quadrilateral $ZRSX$ is cyclic, and similarly, the right angles at S and Y make $XSQY$ cyclic. It follows from the theorem just cited that

$$AR \cdot AZ = AS \cdot AX = AQ \cdot AY,$$

giving

$$AR \cdot AZ = AQ \cdot AY.$$

This implies $ZRQY$ is cyclic. It remains only to show its circumcenter is M.

In the circumcircle of $ZRQY$, the perpendicular bisectors of the chords YQ and ZR intersect at the center. But since YX and QP are perpendicular to YQ, the perpendicular bisector of YQ also passes through the midpoint M of XP. Similarly, the perpendicular bisector of ZR goes through M. Thus M is the point of intersection of these perpendicular bisectors and is indeed the center of the circle. ∎

A somewhat different approach to the pedal circle is given in *More Mathematical Morsels,* Vol. 10, Dolciani Series, page 54.

(ix) The Droz-Farny circles

As usual, let H and O denote the orthocenter and circumcenter of $\triangle ABC$, and let D, E, and F be the feet of the altitudes (figure 87(a)). Now suppose that the circle with center D and radius DO crosses BC at P_1 and P_2, that the circle with center E and radius EO intersects AC at Q_1 and Q_2 and that the circle with center F and radius FO intersects AB at R_1, and R_2.

Then the six points $P_1, P_2, Q_1, Q_2, R_1, R_2$ lie on a circle, and moreover, its center is at H!

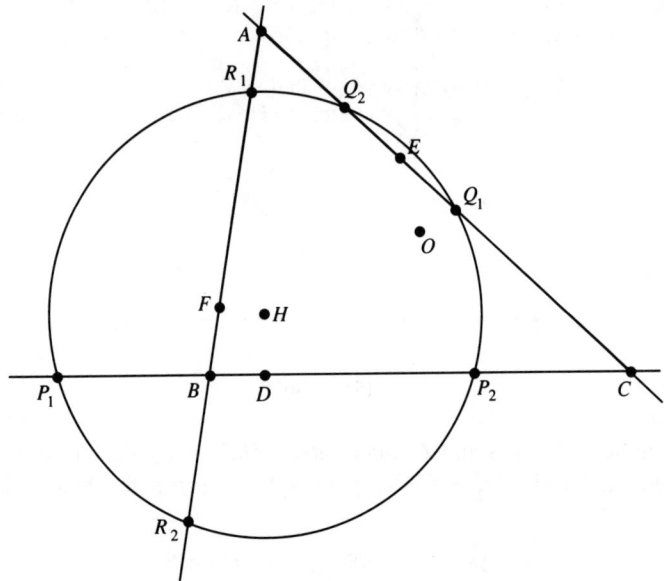

Figure 87(a)

Now suppose we start again, only this time, instead of using the feet of the altitudes for the centers of the circles, let's use the midpoints A', B', and C' of the sides, and instead of having each circle go through O, let each one go through H, i.e. use radii $A'H$, $B'H$, and $C'H$ (Figure 87(b)). Then again

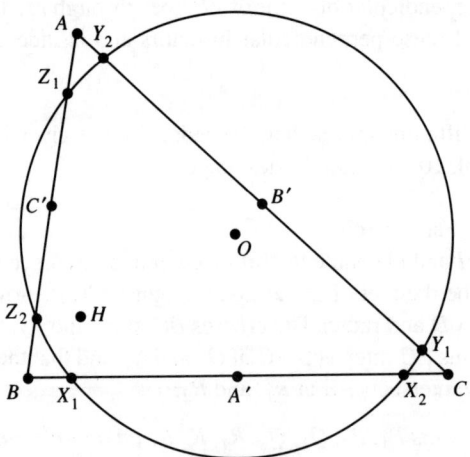

Figure 87(b)

the six points $X_1, X_2, Y_1, Y_2, Z_1, Z_2$ on the sides lie on a circle, this time with center O. On top of this, *these two circles are the same size.*

These are the *Droz-Farny circles* of the pair of points (H, O), named in honor of the nineteenth-century Swiss geometer A. Droz-Farny. Can you guess what is special about the points H and O? It is surprising how easy we will find it to establish the remarkable result that there is a pair of Droz-Farny circles for every pair of isogonal conjugates (P, Q). (It is an easy exercise to prove that H and O are isogonal conjugates of a triangle.)

Suppose, then, that P and Q are any pair of isogonal conjugates of $\triangle ABC$ and that D, E, and F are the feet of the perpendiculars to the sides from one of them, say P, and that circles with centers D, E, and F are drawn to pass through Q. Then the three pairs of points these circles determine on the sides will always lie on a circle with center P, and the two circles that can be constructed in this way are congruent.

The proof is very short and uses the following property of medians (which is easily established by the law of cosines, Figure 88).

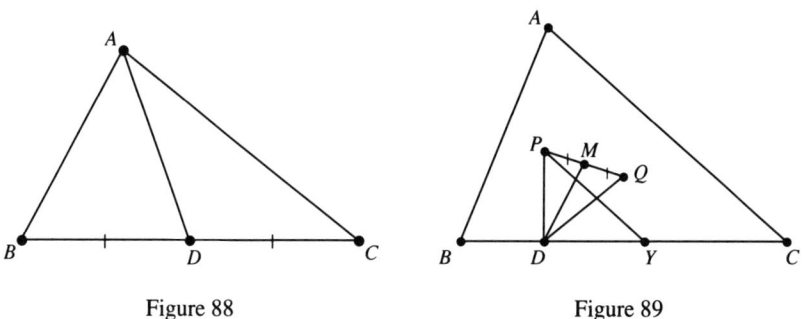

Figure 88   Figure 89

$$AB^2 + AC^2 = 2AD^2 + 2BD^2 = 2AD^2 + 2\left[\frac{1}{2}BC\right]^2$$
$$= 2AD^2 + \frac{1}{2}BC^2.$$

In Figure 89, let the length of PQ be t and let Y be one of the points on BC determined by the circle with center D and radius DQ. Then $DY = DQ$, and recalling that the center of the pedal circle of (P, Q) is the midpoint M of PQ, we have

$$PY^2 = PD^2 + DY^2 \quad \text{(Pythagoras)}$$
$$= PD^2 + DQ^2$$
$$= 2DM^2 + \frac{1}{2}PQ^2 \quad \text{(the median property)}$$
$$= 2\rho^2 + \frac{1}{2}t^2 \quad \text{(where } \rho \text{ is the radius of the pedal circle of } (P,Q)).$$

Since this expression depends only on the distance between the isogonal conjugate points (P, Q) and the radius of their pedal circle, it follows that not only are the six points in question at the same distance from P, implying that they lie on a circle with center P, but that an equal circle with center Q is obtained when the roles of P and Q are interchanged. ∎

We note in passing another remarkable theorem of Droz-Farny.

If two perpendicular straight lines are drawn through the orthocenter H of a triangle, they intercept a segment on each of the sides, and the midpoints of these three segments are collinear!

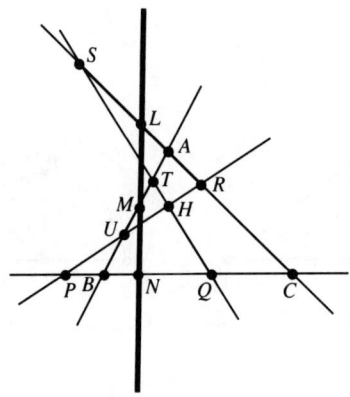

Figure 90

(x) Finally, let us close with another of these dual-personality properties:

The symmedian point of a triangle is the centroid *of its pedal triangle.*

In Figure 91, K is the centroid of $\triangle PQR$.

It suffices to show that PK bisects RQ. Let $KR = x$, $KQ = y$, and the angles at M be α and β, as marked in Figure 91. Then, since $RBPK$ is cyclic (right angles at R and P), exterior angle RKM is equal to the interior and opposite angle B

THE SYMMEDIAN POINT

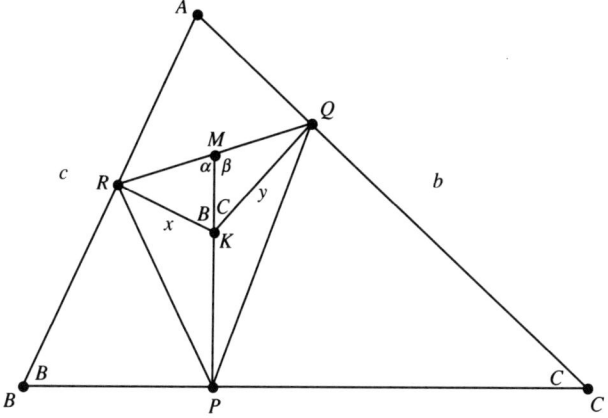

Figure 91

(both are supplements of $\angle PKR$).

Thus $\angle RKM = B$; similarly $\angle QKM = C$.

Hence, in triangles RKM and MKQ, the law of sines gives

$$\frac{RM}{\sin B} = \frac{x}{\sin \alpha} \quad \text{and} \quad \frac{MQ}{\sin C} = \frac{y}{\sin \beta}.$$

But α and β are supplementary, implying $\sin \alpha = \sin \beta$, and dividing the first equation by the second, we get

$$\frac{RM}{MQ} \cdot \frac{\sin C}{\sin B} = \frac{x}{y}, \quad \text{i.e.,} \quad \frac{RM}{MQ} \cdot \frac{c}{b} = \frac{x}{y}.$$

But, for the symmedian point K, we saw that $x/y = c/b$, and therefore $RM/MQ = 1$, implying $RM = MQ$, as desired. ∎

Personally, I am always pleased to use ideas from one field of mathematics in the solution of a problem in another area, as we have just done using the law of sines from trigonometry. However, if you would prefer a purely Euclidean approach, perhaps you would enjoy the following ingenious argument due to John Rigby.

As above, we first establish that the angles B and C occur at K (Figure 92). Next, let RK be extended its own length to S, and let $RK = KS = x$ and $KQ = y$. Then, just as $\angle B$ and $\angle C$ occur at K, so is $\angle A$ present as angle QKS. Also, as in the former proof, because K is a symmedian, we have

$$\frac{x}{y} = \frac{c}{b}.$$

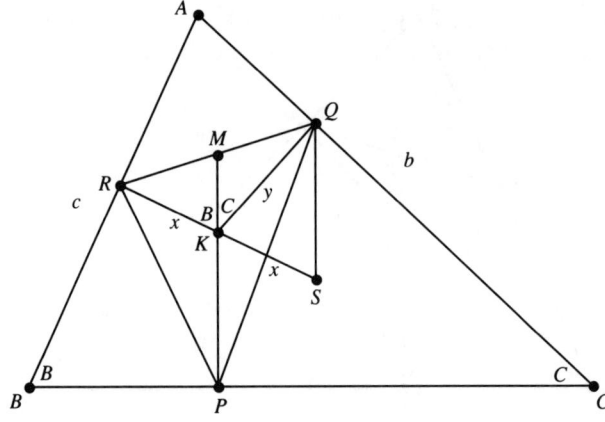

Figure 92

Thus, in triangles ABC and QKS, the sides about equal angles BAC and QKS are proportional, implying the triangles are similar. Thus $\angle KSQ = \angle B$ of $\triangle ABC$, and the equal corresponding angles B at K and S imply that KM and SQ are parallel. Recalling that $RK = KS$, then in $\triangle RSQ$, KM issues from the midpoint K of RS, is parallel to SQ, and therefore bisects the third side RQ, as desired.

References

[1] John Mackay: Early History of the Symmedian Point, Proceedings of the Edinburgh Math. Soc., XI, 1892–93, p. 92.
[2] John Mackay: Symmedians of a Triangle and their Concomitant Circles, Proceedings of the Edinburgh Math. Soc., XIV, 1896, pp. 37–103.

Exercise Set 7

The exercises describe some remarkable properties and are worth reading even if you don't try them. The problems marked with an asterisk are more difficult. While I am quite fond of #6 and #7, I think #8 is a real gem. Solutions are given at the end of the book (where you will find John Rigby's wonderful approach to #'s 7 and 8).

THE SYMMEDIAN POINT

7.1 Let *DEF* be the orthic triangle of $\triangle ABC$, and let K' be the symmedian point of $\triangle AFE$ (Figure 93). Prove that the symmedian AK' of $\triangle AFE$ bisects *BC*.

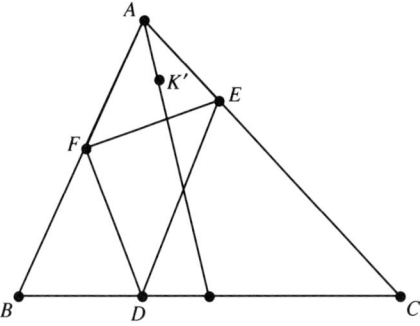

Figure 93

7.2 Let $\alpha = 2\triangle/(a^2 + b^2 + c^2)$, where $\triangle$ is the area of $\triangle ABC$. Prove that the distances *KL*, *KM*, *KN* to the sides of $\triangle ABC$ from the symmedian point *K* are αa, αb, and αc (Figure 94).

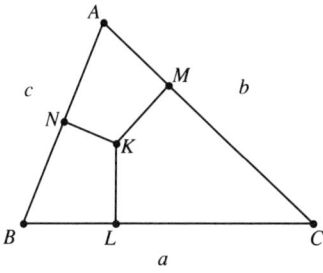 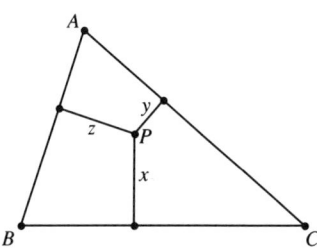

Figure 94   Figure 95

7.3 Prove that the point *P* inside $\triangle ABC$ which minimizes the sum of the squares of the distances to the sides, $S = x^2 + y^2 + z^2$, is the symmedian point *K* (Figure 95). Hint: Lemoine, himself, used the following identity in his proof:

$$(x^2 + y^2 + z^2)(a^2 + b^2 + c^2)$$
$$= (ax + by + cz)^2 + (bx - ay)^2 + (cy - bz)^2 + (cx - az)^2.$$

7.4 Prove that the symmedian point K divides the symmedian AP of $\triangle ABC$ so that

$$\frac{AK}{KP} = \frac{b^2 + c^2}{a^2} \quad \text{(Figure 96)}.$$

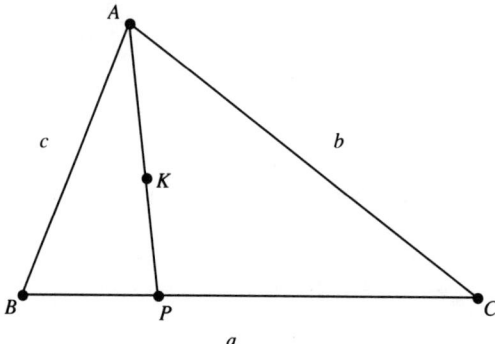

Figure 96

7.5 Prove that the lengths of the sides of the pedal triangle LMN of the symmedian point K of $\triangle ABC$ are proportional to the *medians* of $\triangle ABC$ and that the angles of $\triangle LMN$ are the angles between these medians.

*7.6 Perpendiculars from B and C meet the bisector of angle A at P and Q, respectively (Figure 97). The line from P, parallel to AB, meets the line from Q, parallel to AC, at R. Prove that AR is the symmedian from A.

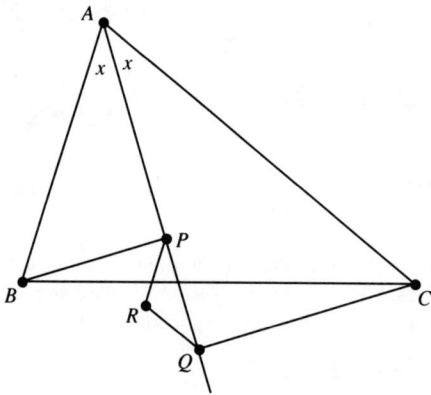

Figure 97

*7.7 The lines AK, BK, and CK, through the symmedian point K of $\triangle ABC$ meet the circumcircle of $\triangle ABC$ at D, E, and F (Figure 98). Prove that K is also the symmedian point of $\triangle DEF$.

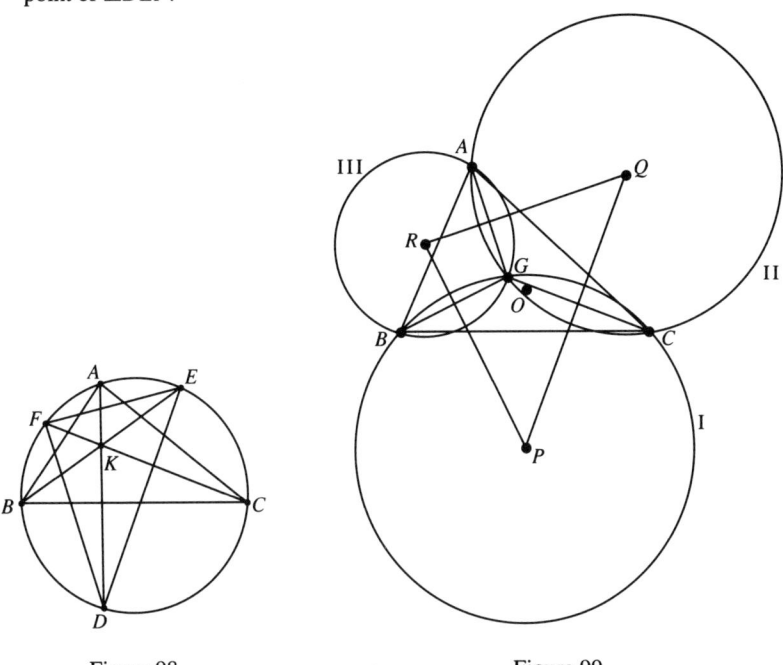

Figure 98

Figure 99

*7.8 Let each of the circles I, II, and III (figure 99) pass through the centroid G of $\triangle ABC$ and two of the vertices. Prove that the circumcenter O and the centroid G of $\triangle ABC$ are, respectively, the centroid and symmedian point of the triangle PQR which is determined by the centers of these circles.

CHAPTER EIGHT

The Miquel Theorem

1. It is not always safe to assume that a historical honor is given in recognition of an original discovery. The Argand plane was discovered by Caspar Wessel, the Simson line by William Wallace, and Wilson's theorem was known to Leibniz and first proved by Lagrange. Occasionally this is simply carelessness, but generally a historical attribution is well merited by significant work in the area, if not by the actual discovery itself. Whether the following theorem originates with Miquel, I don't know. However, in 1838, Miquel explicitly stated and proved it. It is such a simple and engaging result that it is somewhat surprising that it did not come into prominence much earlier.

2. The Theorem of Miquel

Suppose a point is picked at random on each side of $\triangle ABC$ (Figure 100). Then, in every case, the circles AP_3P_2, BP_1P_3, and CP_2P_1 are concurrent.

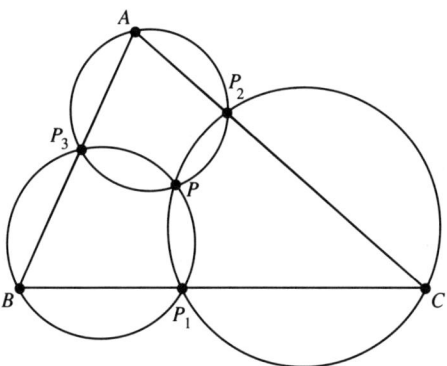

Figure 100

The proof is almost immediate.

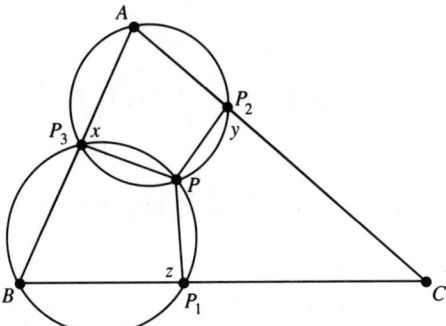

Figure 101

Let P be the second point of intersection of the two circles through A and B. Then, in Figure 101, cyclic quadrilateral AP_3PP_2 gives $x = y$, and P_3BP_1P gives $x = z$, making $y = z$ and PP_1CP_2 cyclic. ∎

Sometimes the point P doesn't lie inside $\triangle ABC$, and in such a case this argument needs a slight adjustment. For example, in Figure 102, AP_3P_2P gives $x = y$ on chord AP, P_3BP_1P gives $y = z$, and then $x = z$; hence $t = 180° - x = 180° - z = s$, making PP_2P_1C cyclic.

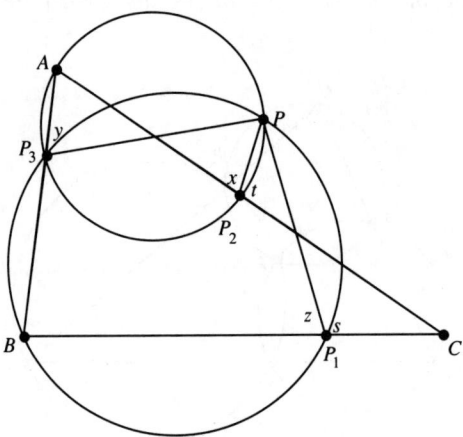

Figure 102

It is really nice, however, that the theorem remains valid even when $P_1, P_2,$ and P_3 are selected anywhere on the sides or their extensions. For example, in Figure 103, in the circle through B, $\angle 2 = \angle 3$ on arc BP, and in the circle through A, $\angle 2 = \angle 1$ on arc AP; hence $\angle 1 = \angle 3$ (on CP), making P_2P_1PC cyclic. ∎

THE MIQUEL THEOREM

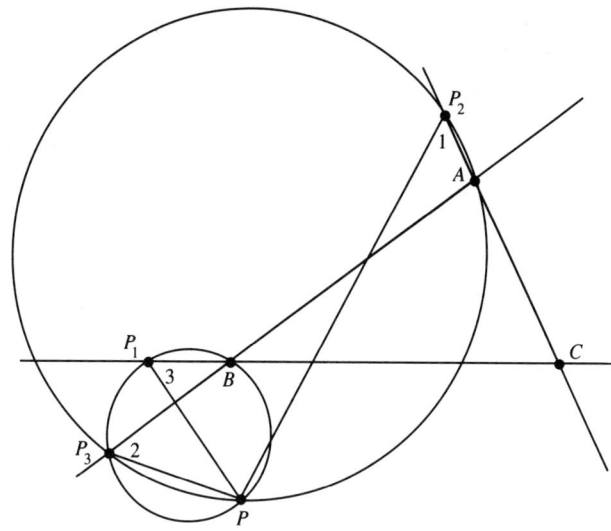

Figure 103

The point P is called the *Miquel point of P_1, P_2, P_3 relative to the triangle ABC*, and the circles involved are called *Miquel circles*. Conversely, the triangle $P_1P_2P_3$ is a Miquel triangle of the point P; we shall see that the same point P gives rise to an infinite family of Miquel triangles.

3. The Case of P_1, P_2, P_3 Collinear

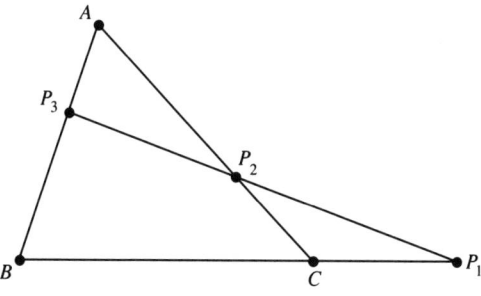

Figure 104

We have seen that the Miquel point P is the second point of intersection of the circles AP_3P_2 and CP_1P_2. If P_1, P_2, P_3 happen to lie on a straight line, let's investigate the situation that arises by considering the basic triangle to be BP_1P_3 and A, P_2, and C as the points chosen on its sides. Again, we find two of the three

Miquel circles are the ones around triangles AP_3P_2 and CP_1P_2, implying that the same Miquel point P is determined. In fact, further investigation reveals that it is always the same Miquel point P that is determined no matter which three of the four straight lines in the configuration are taken to specify the fundamental triangle, with the remaining line giving the points on its sides. Thus we arrive at the remarkable result:

The circumcircles of the 4 triangles formed by any 4 straight lines in general position are concurrent.

Naturally, the point of concurrency P is called *the Miquel point of the quartet of lines*.

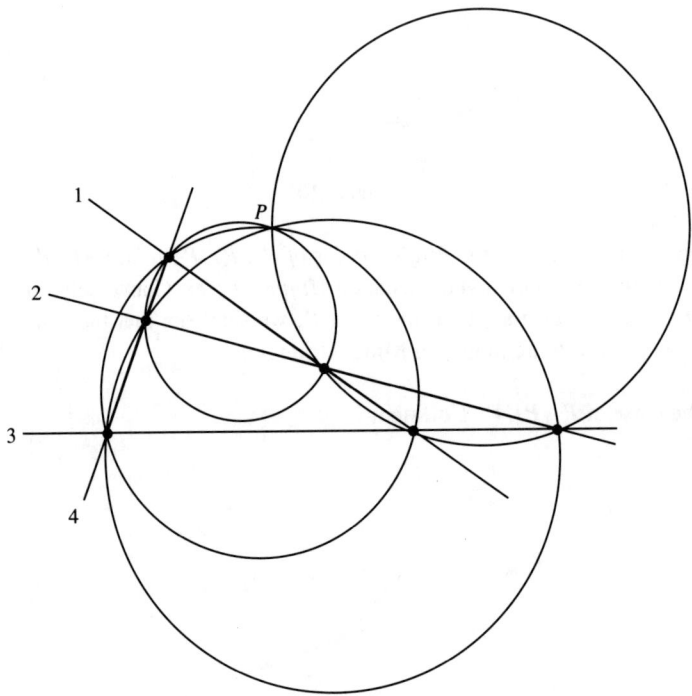

Figure 105

4. Simson Lines

Another neat proof of the concurrence of these four circumcircles is given by the basic property of Simson lines.

Of course, intersecting circles generally have two points of intersection, and it is often necessary to distinguish between them. Let P be the point of intersection of any two of the circumcircles, a point which does not lie on any of the given straight lines themselves—say the circles about the triangles determined by lines 124 and 123 (Figure 106); also, let W, X, Y, and Z be the feet of the perpendiculars from P to the given lines. Then, because P is on the circle around $\triangle 124$, the feet W, X, and Y are collinear on its Simson line; similarly, the feet X, Y, and Z lie on the Simson line of P relative to $\triangle 123$. Since X and Y are common to these lines, they must be the same line, and it is easy to check that $WXYZ$ is the common Simson line of P relative to *any* triangle formed by the four given lines. But because only points on the circumcircle of a triangle have Simson lines, the point P must lie on each of the circumcircles of the four triangles, and the second proof is complete. ∎

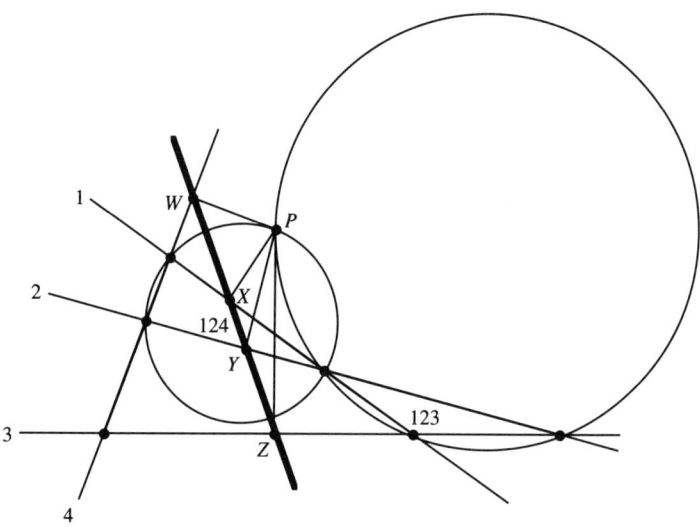

Figure 106

5. A Curious Angle Property

It's always particularly satisfying to come across a nice way of establishing a relation between things that seem to have no apparent connection.

In the original Miquel configuration (Figure 107), the cyclic quadrilaterals immediately imply that the angles marked z are equal. Conversely, it is clear that

if any three lines from P are equally inclined to the sides of $\triangle ABC$ the point P is the Miquel point for the triangle $P_1P_2P_3$ they determine. Thus the same point P is the Miquel point of an infinite family of triangles $P_1P_2P_3$.

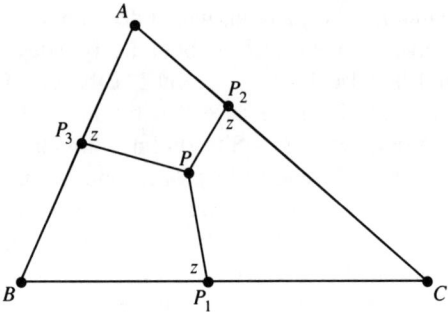

Figure 107

In every case, however, we have the following unexpected relation between the angles:

$$\angle BPC = \angle A + \angle P_3P_1P_2; \quad \text{in figure 108(a),} \quad x = A + y.$$

This is easy to prove as follows.

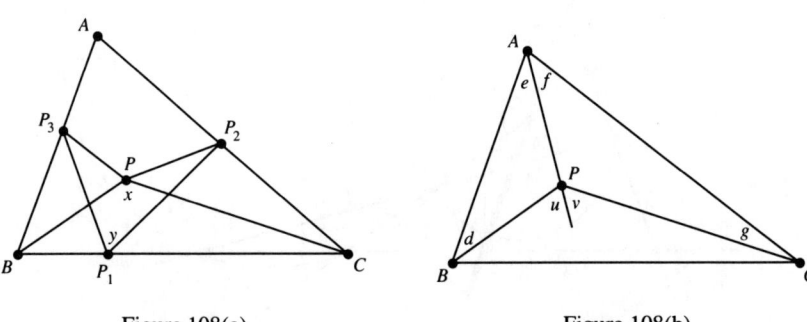

Figure 108(a) Figure 108(b)

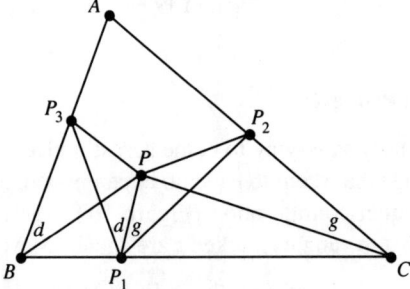

Figure 108(c)

THE MIQUEL THEOREM

Let AP extended split angle x into parts u and v, as shown in figure 108(b). Then, by the exterior angle theorem,

$$u = d + e \quad \text{and} \quad v = f + g,$$

so that

$$x = u + v = d + e + f + g = d + A + g.$$

In Figure 108(c), the cyclic quadrilaterals give equal angles d at B and P_1 and also equal angles g at C and P_1, giving $y = d + g$. Hence $x = A + y$. ∎

Finally, let us close this essay with an alternate approach to this result that I hope you will find of interest.

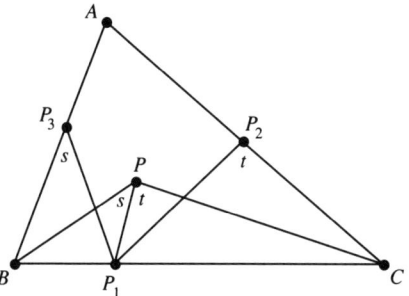

Figure 109

First we join PP_1 and observe that the parts s and t of the angle $x = \angle BPC$ occur again, respectively, in the Miquel circles at P_3 and P_2 (Figure 109).

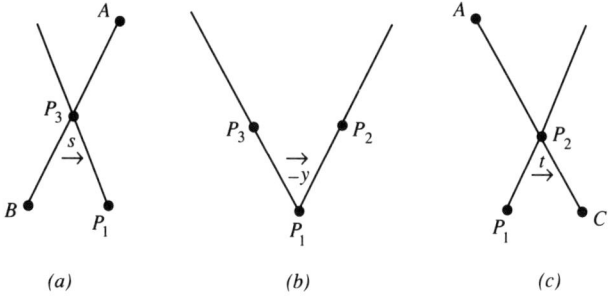

Figure 110

Now let the side AB be rotated about the point P_3 through the positive angle s to make it lie in the direction of P_3P_1 (Figure 110(a)). Next, let's further alter its direction by rotating it about P_1 through the negative angle $-y$ to line it up with P_1P_2 (Figure 110(b)). Finally, let it be spun about P_2 through the positive angle t to lie along the side AC (Figure 110(c)). Thus the net change in direction that has been achieved is just the positive angle A, and we have

$$s - y + t = A,$$

giving

$$x = s + t = A + y. \blacksquare$$

CHAPTER NINE

The Tucker Circles

1. Parallels and antiparallels

There are two ways of choosing points D and E on the sides of $\triangle ABC$ so that $\triangle ADE$ is similar to $\triangle ABC$. In one case, their third sides are parallel, and in the

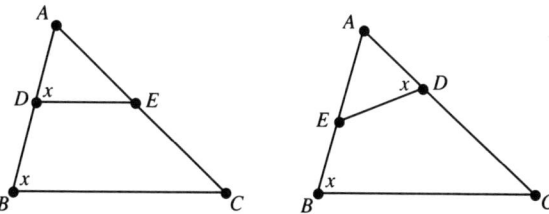

Figure 111

other case their third sides are oblique (Figure 111). In any case, the angles at B and D are always equal, and when DE is not parallel to BC, the quadrilateral $EBCD$ is *cyclic*. Because of this relation to parallels, any line across a triangle which makes $EBCD$ a cyclic quadrilateral is called an *antiparallel* to the opposite side. Thus, in the second figure, ED and BC are antiparallel; in fact, a pair of opposite sides in any cyclic quadrilateral are said to be antiparallel to each other.

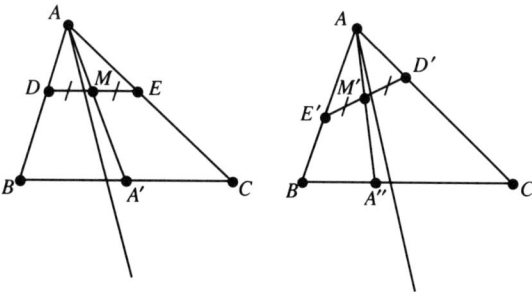

Figure 112

88   EPISODES

Note that, because $\angle ADE$ must equal B, all the antiparallels to BC have the same direction.

It is clear that the median AA' of $\triangle ABC$ *bisects* every parallel DE to BC. Now, if AA' and DE are *reflected in the bisector of angle* A, the parallel turns into an antiparallel and the median into a symmedian (Figure 112). However, the bisection property is not disturbed by this process, and we have the important result that

a median bisects every parallel, and a symmedian bisects every antiparallel.

2. The Lemoine circles

In 1873, the French Association for the Advancement of the Sciences met at Lyons. It was Emile Lemoine's contributions at this meeting that set off the modern revival in Euclidean geometry. He drew attention to some of the properties of the symmedian point K of a triangle and henceforth this point has been known in France and Britain as the *Lemoine point*. Because a symmedian bisects antiparallels, that is, is a *median* of antiparallels, Lemoine referred to the symmedian point as the *center of antiparallel medians*.

Clearly, a parallel to each side of a triangle through a point P inside it determines two points on each side; of course, antiparallels through P do the same. Now generally we are not aware of anything special about the 6 points thus determined on the sides. But if the point P is the symmedian point K, then the 6 points always lie on a circle—parallels give the *first Lemoine circle* and antiparallels the *second Lemoine circle*.

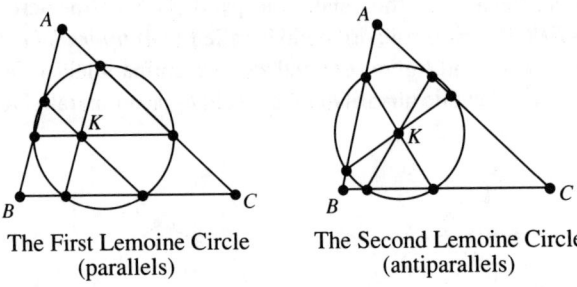

The First Lemoine Circle   The Second Lemoine Circle
(parallels)   (antiparallels)

Figure 113

(a) THE SECOND LEMOINE CIRCLE. It is so easy to establish the second Lemoine circle that we consider it first; we shall treat the first later (Section 5 of this chapter). Since K lies on each symmedian, it bisects each antiparallel through it (symmedians bisect antiparallels). However, the little triangles like KRU (Figure 114) are all isosceles because, due to the cyclic quadrilaterals determined by the antiparallels forming their sides, both the base angles are equal to an angle of the given triangle—in the case of KRU, it is $\angle A$. Hence $KR = KU$, i.e. $x = y$.

THE TUCKER CIRCLES

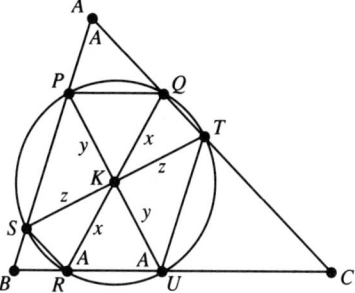

Figure 114

Similarly, in $\triangle KPS$, $y = z$, and we have $x = y = z$. Thus, not only are the 6 points concyclic, but we have the additional fact that *the center of the circle is K itself*.

Now, the vertically opposite angles at K make the isosceles triangles KRU and KQP congruent, and it follows easily that PQ is parallel to BC. Thus the 6 concyclic points lie on an inscribed self-intersecting hexagon $PQRSTU$ whose sides are alternately parallel and antiparallel to the sides of ABC:

PQ, RS, and TU are parallel to the sides, while QR, ST, and UP are antiparallel to the sides.

3. The Tucker circles

The remarkable thing is that such an inscribed hexagon, *begun at any point P on a side*, always yields a *closed* hexagon which is *cyclic*; the circle thus determined is called a *Tucker Circle*. Thus the second Lemoine circle is clearly

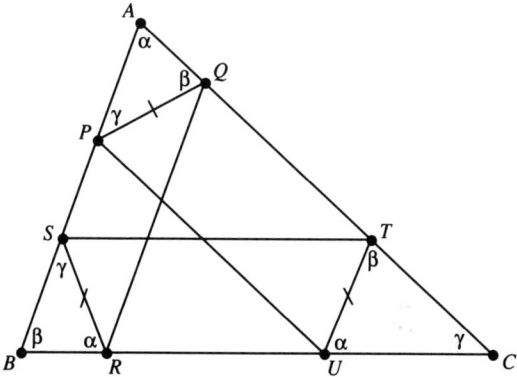

Figure 115

a Tucker circle. We shall see that the first Lemoine circle is also a Tucker circle, but first let's justify these amazing general claims that have just been made.

First of all, let's show that going around a triangle, starting at any point on a side, and alternately drawing parallels and antiparallels in fact always yields a *closed* hexagon. It is only a matter of showing that, after drawing the fifth side, the segment which closes the hexagon is the third parallel or antiparallel as the case may be. Let us begin with the case in which the first edge drawn is an *antiparallel* (Figure 115). Accordingly, let points P, Q, R, S, T, U be determined by 3 antiparallels PQ, RS, TU, and 2 parallels QR and ST. We need to show that UP is parallel to AC.

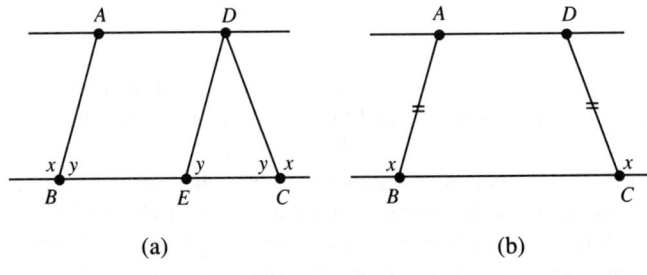

Figure 116

We can do this nicely by recalling two basic properties of parallel lines:

(i) a pair of transversals AB and CD which are equally inclined to a pair of parallel lines are themselves equal: $AB = CD$ (Figure 116(a));

(ii) conversely, if AB and CD are equal segments on the same side of a line BC and are equally inclined to it, then their endpoints A and D determine a line parallel to BC: $AD \parallel BC$ (Figure 116(b)).

In Figure 115, the antiparallel PQ makes $\angle APQ = \gamma \ (= \angle C)$ and the antiparallel RS makes $\angle BSR = \gamma$. Hence PQ and RS are equally inclined to the parallels PS and QR, and we conclude that $PQ = RS$. Similarly $RS = UT$, and we have $PQ = RS = UT$.

Because PQ is antiparallel to BC, we have $\angle AQP = \angle \beta \ (= \angle B)$; similarly, antiparallel UT makes $\angle UTC = \angle \beta$. Hence PQ and UT are not only equal but are equally inclined to QT, and so it follows that UP is parallel to QT, i.e., to AC. ∎

Thus we have shown that a "Tucker hexagon" whose first edge is an antiparallel does in fact close with the third parallel. To settle this matter completely, we still need to show that a closed Tucker hexagon results when the first side is a parallel rather than an antiparallel. Accordingly, suppose we start a Tucker hexagon

UPQRST with *UP* parallel to *AC*—then *UP*, *QR* and *ST* are parallels, and *PQ* and *RS* are antiparallels (Figure 117). It is easy to show that *TU* is antiparallel to *AB*.

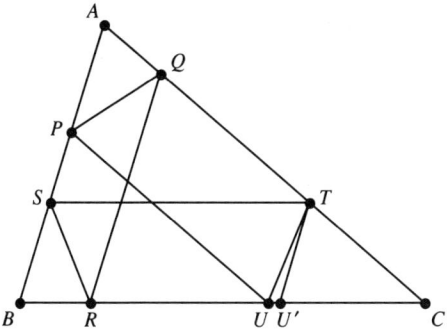

Figure 117

Let TU' be antiparallel to *AB* and consider the Tucker hexagon $PQRSTU'$ that starts with the antiparallel *PQ*. We have just proved that such a hexagon has its closing side $U'P$ parallel to *AC*. But *UP* is given parallel to *AB*, so *PU* and PU' are themselves parallel, making $U = U'$. Thus *TU* is indeed antiparallel to *AC*. ∎

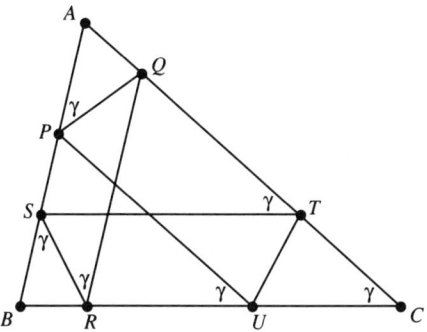

Figure 118

Now let's address the major claim that *a Tucker hexagon PQRSTU is cyclic*. In Figure 118, the angle γ occurs at each of the six places marked—*at P and S because PQ and RS are antiparallels, and elsewhere because of the parallel lines*. Since a quadrilateral is cyclic if the exterior angle at one vertex is equal to the interior angle at the opposite vertex, the angles γ at *S* and *U* show that *PSRU* is cyclic, and the angles γ at *P* and *R* show that *PSRQ* is cyclic. Since *P*, *S*, and *R* are on the circumcircles of these quadrilaterals, they must be the same circle and hence the five points *Q*, *P*, *S*, *R*, and *U* all lie on a circle *Z*. The angles γ at

92   EPISODES

P and *T* show that *PSTQ* is cyclic, and therefore *T* also lies on this circle *Z* that goes through *Q, P*, and *S* and the conclusion follows. ∎

Since a hexagon may be started at any point on a side, the Tucker hexagons of a triangle range over a wide assortment of shapes and sizes. It strikes me as remarkable, however, that the *center* of every Tucker circle lies on the line *KO* through the symmedian point *K* and the circumcenter *O* of the given triangle *ABC*. Let us consider next the following really nice proof of this intriguing result.

4. The center of a Tucker circle lies on the line *KO*

In Figure 119, let the Tucker hexagon *PQRSTU* begin with antiparallel *PQ*. Then, if *QR* meets *PU* at *X*, quadrilateral *APXQ* has opposite sides parallel and is therefore a parallelogram. Consequently, *AX* bisects *PQ* at *L*. Similarly, *BY* bisects *RS* at *M* and *CZ* bisects *UT* at *N*. We have already seen that the antiparallels of a Tucker hexagon are equal, and so it follows that the six halves are also equal:

$$PL = LQ = SM = MR = UN = NT.$$

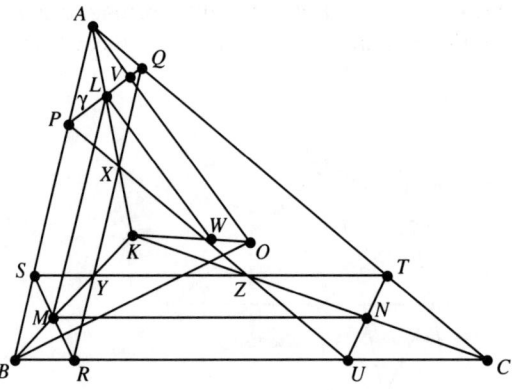

Figure 119

Now, because a symmedian bisects an antiparallel, *ALX* is the symmedian from *A*. Similarly, *BY* and *CZ* are symmedians and so *AX*, *BY*, and *CZ* meet at the symmedian point *K*.

Because antiparallels *PQ* and *RS* are equal and *PS* and *QR* are parallel, the line *LM* runs down the middle of trapezoid *PSRQ* and is therefore parallel to *PS*, and hence to *AB*. Similarly, *MN* is parallel to *BC*. Thus *L* divides *AK* in the same ratio that *M* divides *BK* which, in turn, is the same as the ratio in which *N* divides *CK*. Letting $KL/KA = \mu$, the dilatation $K(\mu)$ takes $\triangle ABC$ and its circumcenter *O* to

$\triangle LMN$ and its circumcenter W, where W lies on KO so that the corresponding radii OA and WL are parallel. As we shall see next, OA is perpendicular to PQ, and hence it follows that WL is also perpendicular to PQ.

We have noted above that $\angle APQ = \gamma = \angle ACB$. Also, in the circumcircle of $\triangle ABC$, the chord AB subtends at the center an angle that is twice the angle it subtends at C on the circumference, and we have

$$\angle AOB = 2\gamma.$$

Accordingly, in isosceles triangle AOB, the base angle

$$\angle OAB = \frac{1}{2}(180° - 2\gamma) = 90° - \gamma.$$

In $\triangle APV$, then, the angle at V is a right angle, making OA perpendicular to PQ.

Similarly, the radii WM and WN are respectively perpendicular to their antiparallels RS and TU (Figure 120). Thus, at each of L, M, and N, there are two right-triangles, all six of which are congruent (two sides and the included angle; recall W is the circumcenter of $\triangle LMN$, making WL, WM, and WN equal radii), and each of P, Q, R, S, T, and U is at the same distance from W, making W the center of the Tucker circle in question. ∎

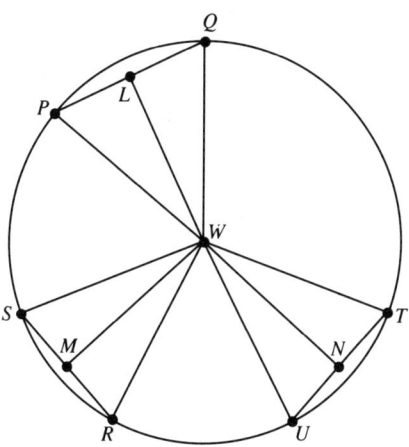

Figure 120

Finally, we note that, as in Figure 119, when the antiparallel PQ crosses KA between K and A, the center W lies *between* K and O. When L divides KA externally, however, W does the same to KO (Figure 121(a) and (b)).

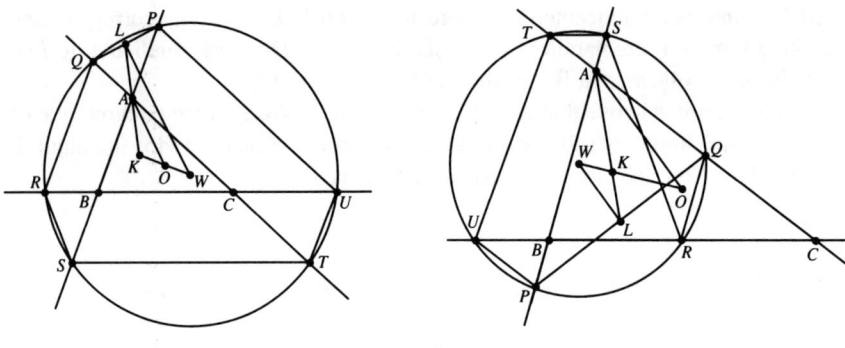

Figure 121(a) Figure 121(b)

5. The first Lemoine circle

Now let's deal with the first Lemoine circle, determined by parallels to the sides of $\triangle ABC$ through its symmedian point K. Referring to Figure 122(a), we need to show that the non-parallel sides PQ, RS, and TU of the defining hexagon are in fact antiparallels.

Consider the case of PQ. Having opposite sides parallel, $APKQ$ is a parallelogram, making L the midpoint of PQ. But we know that symmedians bisect antiparallels, and the following easy argument shows that PQ must be an antiparallel:

if the antiparallel through P were a different segment PY (Figure 122(b)), then, with PY bisected at X by symmedian AK, we would have LX parallel to QY in $\triangle PQY$; but LX and QY meet at A.

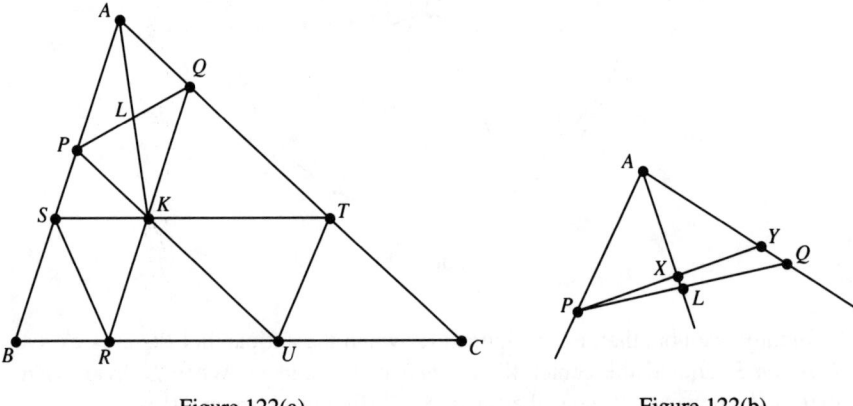

Figure 122(a) Figure 122(b)

Thus the first Lemoine circle is a Tucker circle. We observe that the ratio

$$\mu = \frac{KL}{KA} = \frac{1}{2},$$

so the center W is the midpoint of the segment KO connecting the symmedian point K with the circumcenter O of $\triangle ABC$. In summary, then, we have:

> In $\triangle ABC$, the center W of the first Lemoine circle is the midpoint of KO, and the center of the second Lemoine circle is the circumcenter O of $\triangle ABC$.

Accordingly, we can view the circumcircle as the Tucker circle in which P, Q, and A all coincide.

6. The Taylor Circle

Let's conclude our look at the family of Tucker circles with another of its distinguished members.

From the foot of each altitude perpendiculars are drawn to each of the other two sides (Figure 123). Then the 6 feet of these perpendiculars determine the *Taylor Circle* of $\triangle ABC$. In order to establish this circle, we need only show that the six feet are the vertices of a Tucker hexagon of parallels and antiparallels, which we know is always cyclic.

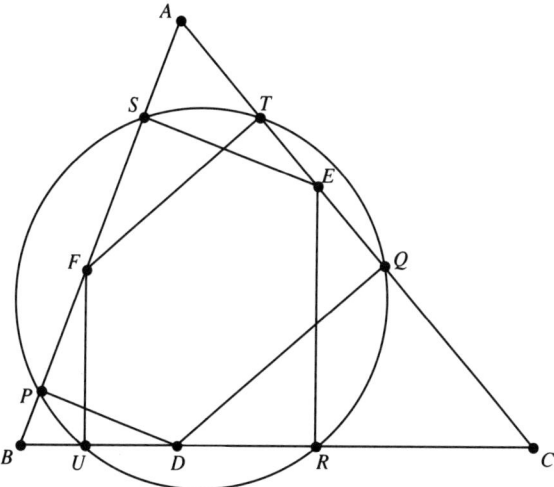

Figure 123

It is easy to show that the sides PQ, RS, and TU are antiparallels. In the case of PQ, for example (Figure 124(a)), we quickly obtain $\angle APQ = \gamma (= \angle C)$ as follows:

The right angles at P and Q make $APDQ$ cyclic, in which $\angle APQ = \angle ADQ$ on chord AQ; since right triangles ADQ and ADC have $\angle DAQ$ in common, their third angles are equal, giving $\angle ADQ = \gamma$. Hence, $\angle APQ = \gamma$, making PQ the desired antiparallel.

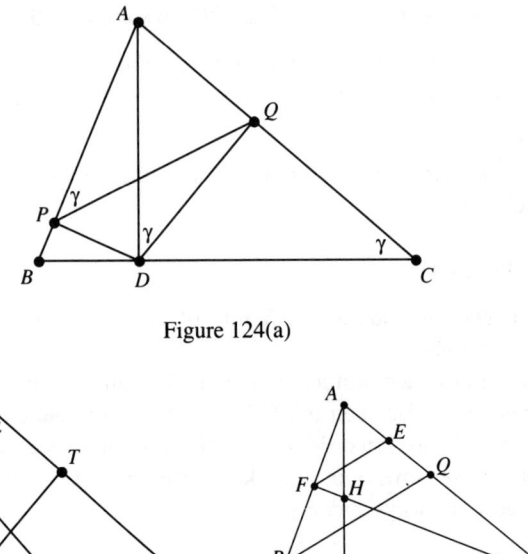

Figure 124(a)

Figure 124(b)

Figure 124(c)

The difficulty is to prove that the other three sides are parallel to the sides of $\triangle ABC$.

Perhaps surprisingly, one approach to this is to stay with the antiparallels and show that they all have the same length. In figure 124(b), it is clear that the antiparallels RS and TU each determine the angle α at R and U, showing that they are equally inclined to BC. Thus, if they are equal in length as well, ST is parallel to BC, and our argument is complete. Consequently, let us determine the length of the antiparallels, beginning with PQ.

In Figure 124(c), since EF is a side of the orthic triangle, it is also an antiparallel to BC. Thus $FE \parallel PQ$ (all antiparallels to BC are parallel), and we have

$$\frac{PQ}{FE} = \frac{AP}{AF}.$$

Now clearly $PD \parallel FH$, giving

$$\frac{AP}{AF} = \frac{AD}{AH}.$$

Hence

$$\frac{PQ}{FE} = \frac{AD}{AH}, \quad \text{from which} \quad PQ = AD\left(\frac{EF}{AH}\right).$$

Now let us pursue the value of EF/AH. In Figure 125, the right angles at E and F make $AFHE$ cyclic, giving equal angles x at F and H. In $\triangle AFE$, the law of sines yields

$$\frac{EF}{\sin A} = \frac{AE}{\sin x}.$$

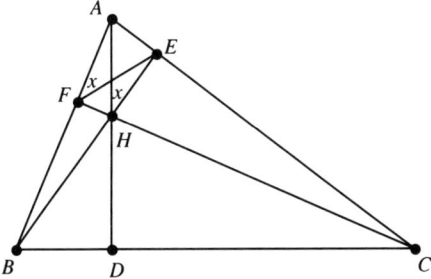

Figure 125

From right triangle AEH,

$$\sin x = \frac{AE}{AH},$$

and so

$$\frac{EF}{\sin A} = \frac{AE}{\sin x} = AH, \quad \text{from which} \quad \frac{EF}{AH} = \sin A.$$

Thus $PQ = AD \cdot \sin A$.

But $\frac{1}{2} \cdot AD \cdot a = (\text{Area } \triangle ABC) = T$, say, and we obtain

$$AD = \frac{2T}{a},$$

giving

$$PQ = 2T \cdot \frac{\sin A}{a}.$$

Similarly, the other two antiparallels have lengths

$$RS = 2T \cdot \frac{\sin B}{b} \quad \text{and} \quad TU = 2T \cdot \frac{\sin C}{c},$$

which, by the law of sines, are equal to PQ. ∎

Exercise Set 9

9.1 Prove that every antiparallel to BC in $\triangle ABC$ is parallel to the tangent to the circumcircle of $\triangle ABC$ at A.

9.2 In Chapter 7 on the symmedian point, we encountered the Gergonne point G, the Gergonne triangle DEF, and the Adams circle $PQRSTU$ of a triangle ABC (see Figure 77 on page 62). In Figure 126, suppose the segments QR, ST, and UP are extended in both directions to form $\triangle XYZ$. Prove that the Gergonne point G is the symmedian point of $\triangle XYZ$ and that the Adams circle of $\triangle ABC$ is the first Lemoine circle of $\triangle XYZ$.

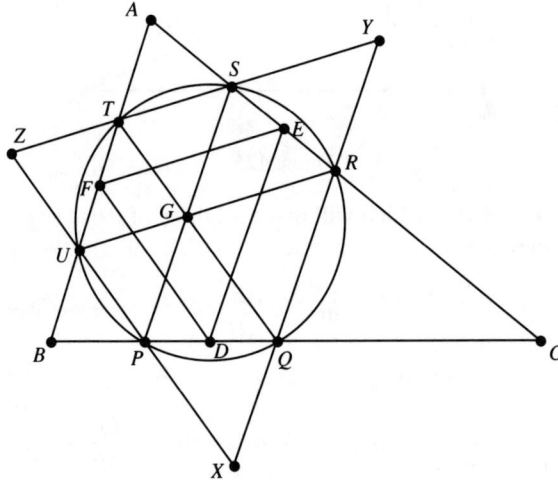

Figure 126

CHAPTER TEN

The Brocard Points

1. The Brocard Points

If a point P inside a triangle ABC is joined to each vertex, three angles x, y, z are formed with the sides (Figure 127). In most cases, these angles are all

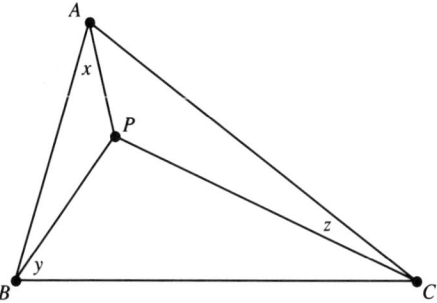

Figure 127

different. Of course, we could always arrange to have two of them the same by simply making y a copy of x. Even if this were done, it would be expecting a lot to look for the same angle again at z. You might well wonder whether there exists a point P inside $\triangle ABC$ which will make all three of these angles equal. As we shall see, there is in fact precisely one such point, often denoted by Ω (omega). However, there is also a unique companion point, Ω', with the same property concerning the alternate triple of angles on the other sides of the segments to the vertices (Figure 128). These remarkable points are called the *Brocard points* of $\triangle ABC$ in honor of Henri Brocard (1845–1922), a French army officer who focussed attention on them in 1875. Brocard was not the first to discover them, for they were known to Crelle, Jacobi, and others some 60 years earlier; however, their work in this area was neglected and soon forgotten.

In order to make equal angles ω at A and B, the circle through A, Ω, and B must be tangent to BC at B (Figure 129(a)). Similarly, the angles ω at A and C

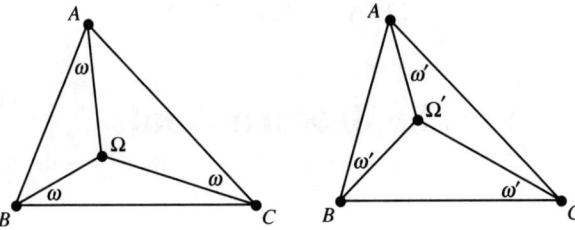

Figure 128

require the circle through A, Ω, and C to be tangent to AB at A. Therefore the only possible location for the point Ω is the point where these circles cross inside the triangle.

Similarly, the companion point Ω' is the intersection of the circle through A and C which touches BC at C and the circle through A and B which touches AC at A. Let's distinguish between Ω and Ω' by calling Ω the *first* Brocard point and Ω' the *second* Brocard point of $\triangle ABC$.

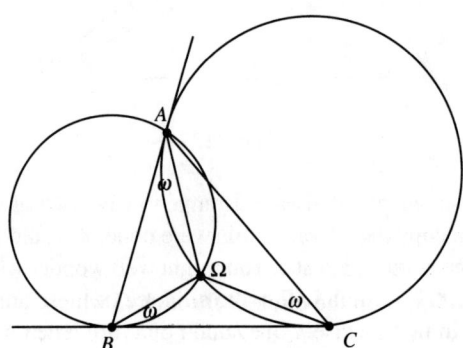

Figure 129(a)

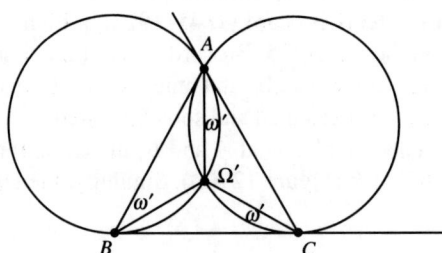

Figure 129(b)

2. The Brocard Angle

Inquiring where the isogonal conjugate W of Ω is (see Chapter 7 on the symmedian point), we immediately find $W = \Omega'$ (Figure 130): since AW is the isogonal conjugate of $A\Omega$, it must make the same angle ω with AC that $A\Omega$ makes with AB; similarly for BW and CW, and since Ω' is the *only* point with this equal-angle property, then $W \equiv \Omega'$.

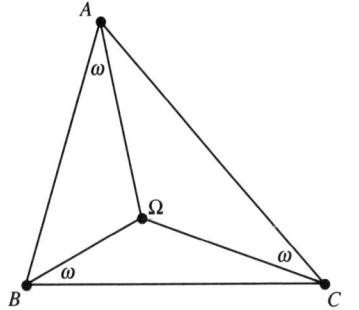 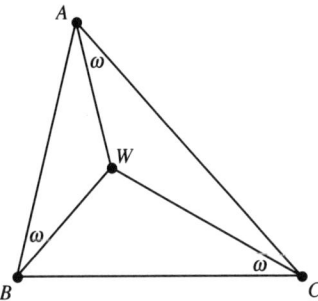

Figure 130

This also gives us the corollary that the Brocard angle ω', determined by Ω', is also ω, and so there is really only one Brocard angle ω. The size of ω can be calculated from the formula

$$\cot \omega = \cot A + \cot B + \cot C,$$

which is easily obtained from the following alternative construction for the Brocard point Ω.

We know that Ω lies on the circle Z through C which touches AB at A. If the line through A parallel to BC meets Z at P, then PB intersects Z at the Brocard point Ω (Figure 131(a)):

each of the pertinent angles at A, B, and C is simply the angle $x = \angle APB$:

- $\angle \Omega AB$ because AB is a tangent and $A\Omega$ a chord;
- $\angle \Omega BC$ because $AP \parallel BC$; and
- $\angle \Omega CA$ is in the same segment of Z as $\angle APB$.

Now, let PE be drawn perpendicular to BC and let AD be the altitude from A (Figure 131(b)). Then $ADEP$ is a rectangle and $AD = PE$. Because AB is a tangent to Z, we have $\angle A = \angle BAC = \angle APC$, and with the parallel lines, the equal alternate angles at P and C give $\angle PCE = \angle A$. Since $\angle PBC = \omega$, we have

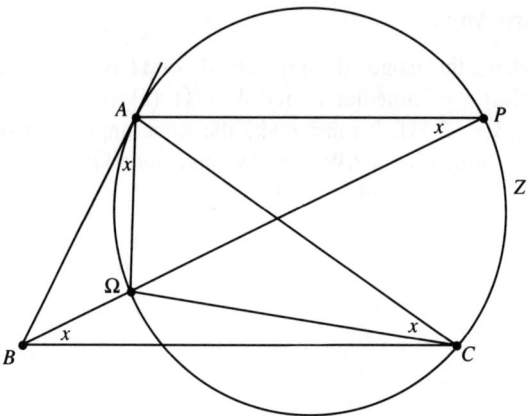

Figure 131(a)

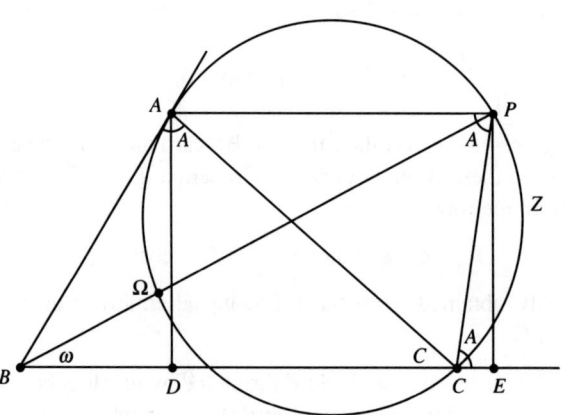

Figure 131(b)

$$\cot \omega = \frac{BE}{PE} = \frac{BD + DC + CE}{PE}$$
$$= \frac{BD}{AD} + \frac{DC}{AD} + \frac{CE}{PE} \quad \text{(recall } PE = AD\text{)}$$
$$= \cot B + \cot C + \cot A. \blacksquare$$

THE MAXIMUM BROCARD ANGLE.

From the formula for cot ω we can deduce the engaging result that

the greatest Brocard angle is 30°.

THE BROCARD POINTS

This follows from the fact that (as we prove below), for every triangle ABC,

$$\cot A + \cot B + \cot C \geq \sqrt{3}.$$

Because $\cot 30° = \sqrt{3}$, then

$$\cot \omega \geq \cot 30°;$$

and since the cotangent function decreases in $(0, \pi)$, this yields

$$\omega \leq 30°.$$

We may establish the above inequality as follows: Since $A + B + C = 180°$,

$$\begin{aligned}
S &= \cot A + \cot B + \cot C \\
&= \cot A + \cot B + \cot[180 - (A + B)] \\
&= \cot A + \cot B - \cot(A + B) \\
&= \cot A + \cot B - \frac{\cot A \cdot \cot B - 1}{\cot A + \cot B}.
\end{aligned}$$

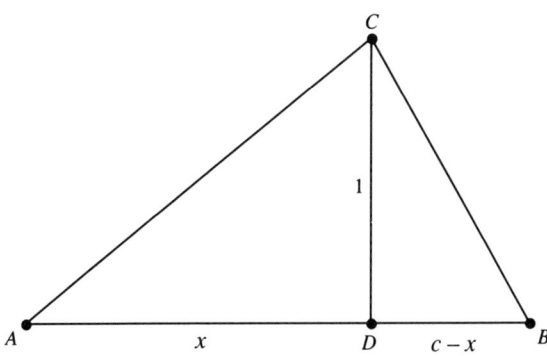

Figure 132

Now, without loss of generality we can label the angles of $\triangle ABC$ so that $A \leq B \leq C$. In this case, the altitude CD (from a biggest angle) will meet the opposite side at an *interior* point D, and for some positive x, we have $AD = x$ and $DB = c - x$ (Figure 132). Choosing CD to be the unit of length, we have

$$\cot A = x, \quad \cot B = c - x,$$

making

$$\cot A + \cot B = c.$$

Thus

$$S = c - \frac{x(c - x) - 1}{c},$$

so that
$$cS = c^2 - xc + x^2 + 1,$$
which, by an ingenious rearrangement, gives
$$cS = \left(x - \frac{c}{2}\right)^2 + \left(\frac{\sqrt{3}}{2}c - 1\right)^2 + \sqrt{3}\,c \geq \sqrt{3}\,c.$$
Hence $S \geq \sqrt{3}$, with equality if and only if
$$c = \frac{2}{\sqrt{3}} \quad \text{and} \quad x = \frac{c}{2} = \frac{1}{\sqrt{3}}.$$
For an equilateral triangle, then,
$$\cot \omega = 3 \cdot \cot 60° = 3 \cdot \frac{1}{\sqrt{3}} = \sqrt{3},$$
showing that ω is a maximum in this case. ∎

Incidentally, the triangle inequality $a^2 + b^2 + c^2 \geq 4\sqrt{3}\,T$, where T denotes the area of $\triangle ABC$, known as Weitzenböck's inequality, follows readily from our discussion: with altitude $CD = 1$ (Figure 132)
$$T = \text{Area}(\triangle ABC) = \frac{1}{2} \cdot c \cdot 1 = \frac{1}{2}c;$$
and, applying the theorem of Pythagoras to triangles ADC and CDB, we have
$$\begin{aligned} a^2 + b^2 + c^2 &= [(c-x)^2 + 1] + (x^2 + 1) + c^2 \\ &= 2(c^2 - xc + x^2 + 1) \\ &= 2cS \\ &\geq 2c\sqrt{3} \\ &= 4\sqrt{3}\,T, \quad \text{since } T = \frac{1}{2}c. \blacksquare \end{aligned}$$

Exercise
Show that
$$\cot \omega = \frac{a^2 + b^2 + c^2}{4T}.$$

A RELATION BETWEEN Ω AND Ω'.

Let the extensions of lines $A\Omega$, $B\Omega$, and $C\Omega$ intersect the circumcircle of $\triangle ABC$ at B', C', A' (Figure 133(a)). Then surprisingly,

$\triangle A'B'C'$ is congruent to $\triangle ABC$.

THE BROCARD POINTS

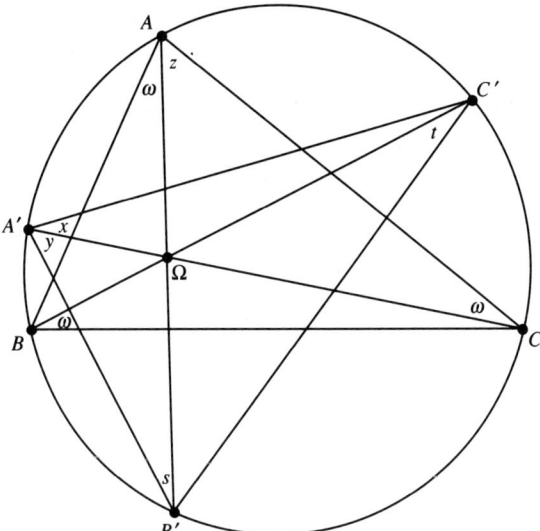

Figure 133(a)

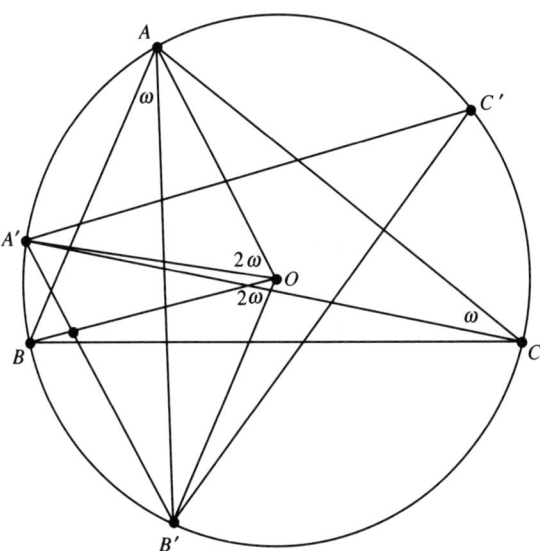

Figure 133(b)

To see this, note that the angle x at A', standing on arc CC', is equal to the angle ω at B. Similarly, the angle y at A', on arc $B'C$, is equal to z at A, and we have

$$\angle A' = x + y = \omega + z = A.$$

Now equal angles in a circle can only be subtended by equal chords, and therefore $B'C' = BC$. Similarly for the other sides, and the triangles are congruent. ∎

We have noted that angle x equals the ω at B; similarly s at B' equals the ω at C and t at C' equals the ω at A. Thus $x = s = t = \omega$, and we observe that the Brocard point Ω of $\triangle ABC$ occurs in the position of the *second* Brocard point in $\triangle A'B'C'$.

Since a central angle of a circle is twice the size of a peripheral angle intercepting the same arc, we have (Figure 133(b))

$$\angle AOA' = 2 \cdot \angle ACA' = 2\omega,$$

and similarly $\angle BOB' = \angle COC' = 2\omega$. Thus a rotation $O(2\omega)$ in the appropriate direction takes $\triangle A'B'C'$ into coincidence with $\triangle ABC$, and thereby takes the second Brocard point (Ω) of $\triangle A'B'C'$ to the second Brocard point (Ω') of $\triangle ABC$. Thus

(i) Ω and Ω' are equidistant from the circumcenter O: $\Omega O = \Omega' O$, and
(ii) $\angle \Omega O \Omega' = 2\omega$ (Figure 134).

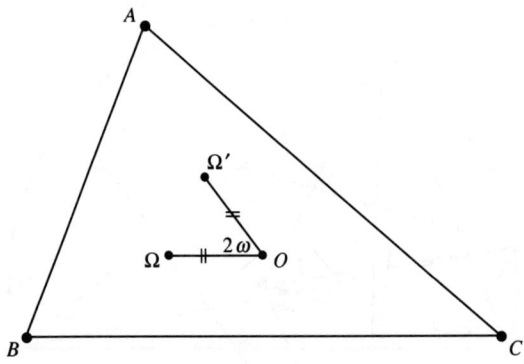

Figure 134

Exercise

Prove that a Brocard point Ω of a triangle is also a Brocard point of the pedal triangle of Ω.

3. The Brocard Circle

There is nothing on the surface of things to suggest any connection between the Brocard points and the *symmedian point* K. Remarkably, however, *the circle on diameter KO always goes through both Brocard points,* and is appropriately called the *Brocard circle* of $\triangle ABC$ (Figure 135).

THE BROCARD POINTS

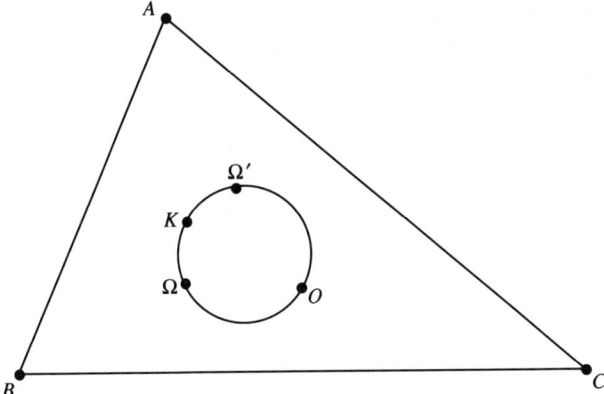

Figure 135

To prove this property we need the results of two of the exercises given earlier.

(i) Recall that K is the unique point in the triangle whose distances to the sides are proportional to the sides themselves; that is, in Figure 136, for some number α,

$$x = \alpha a \qquad y = \alpha b \qquad z = \alpha c.$$

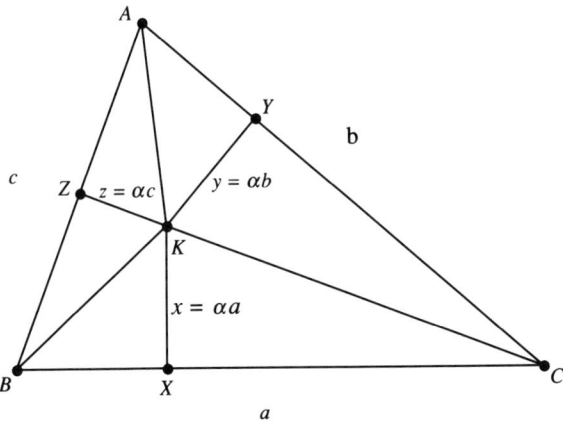

Figure 136

Let us begin by showing that

$$\alpha = \frac{2\Delta}{a^2 + b^2 + c^2},$$

as claimed in Exercise 2 of Set 7 (page 75). Clearly *KA*, *KB*, and *KC* partition $\triangle ABC$ into three triangles whose areas add up to the area $\triangle$ of triangle *ABC*, and we have

$$\frac{1}{2}a(\alpha a) + \frac{1}{2}b(\alpha b) + \frac{1}{2}c(\alpha c) = \frac{1}{2}\alpha(a^2 + b^2 + c^2) = \triangle,$$

from which the desired value of α follows immediately.

(ii) Next, let's tie this in with the Brocard angle by showing that

$$\cot \omega = \frac{a^2 + b^2 + c^2}{4\triangle},$$

as required in the exercise on page 104. In Figure 137, let *h* be the length of the altitude *AD*. Then

$$a = BC = BD + DC$$
$$= h \cot B + h \cot C$$
$$= h(\cot B + \cot C),$$

and hence

$$a^2 = a \cdot a = ah(\cot B + \cot C).$$

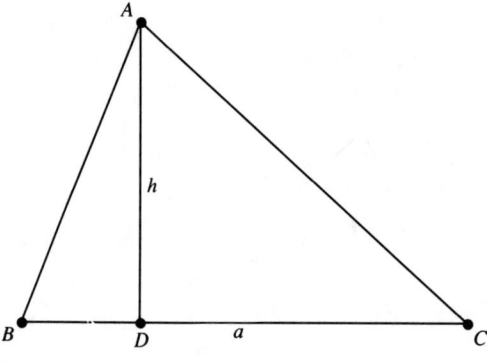

Figure 137

Since the area $\triangle = \frac{1}{2}ah$, we can substitute $2\triangle$ for *ah* obtaining

$$a^2 = 2\triangle(\cot B + \cot C).$$

Similarly,

$$b^2 = 2\triangle(\cot C + \cot A),$$

and
$$c^2 = 2\triangle(\cot A + \cot B).$$

Adding, we get
$$a^2 + b^2 + c^2 = 4\triangle(\cot A + \cot B + \cot C),$$

and since
$$\cot \omega = \cot A + \cot B + \cot C,$$

we obtain
$$\cot \omega = \frac{a^2 + b^2 + c^2}{4\triangle}. \blacksquare$$

Thus the constant of proportionality α is given by
$$\alpha = \frac{2\triangle}{a^2 + b^2 + c^2} = \frac{1}{2}\tan \omega,$$

and the distances αa, αb, and αc from the symmedian point K to the sides of $\triangle ABC$ are respectively
$$\frac{1}{2}a \tan \omega, \quad \frac{1}{2}b \tan \omega, \quad \text{and} \quad \frac{1}{2}c \tan \omega.$$

(iii) Finally, let's connect this with the Brocard circle.

As usual, let A' denote the midpoint of BC, and suppose that $B\Omega$ meets $A'O$ at P (Figure 138). Then $BA' = \frac{1}{2}a$ and $\angle PBA' = \omega$, and so

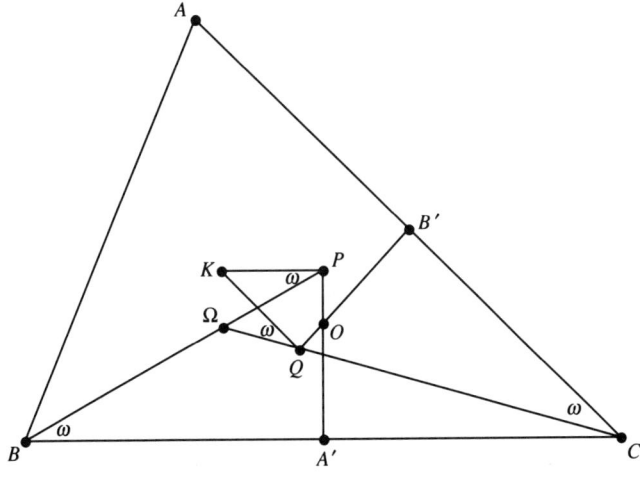

Figure 138

$$PA' = BA' \tan \omega = \frac{1}{2} a \tan \omega,$$

and we conclude that P and the symmedian point K are the same distance above BC. Thus KP is parallel to BC, and since $A'O$ is the perpendicular bisector of BC, it follows that $\angle KPO$ is a right angle.

Similarly, if Q is the point of intersection of ΩC and the perpendicular bisector $B'O$ of AC, then KQ is parallel to AC and $\angle KQO$ is also a right angle. Therefore the Brocard circle, on KO as diameter, goes through both P and Q.

Now, because $KP \parallel BC$, and similarly $KQ \parallel CA$, we have equal alternate angles ω at B and P, and equal corresponding angles ω at Q and C, showing that ΩK subtends equal angles ω at P and Q and implying that the circle through K, P, and Q also goes through Ω. But we have just seen that the circle through K, P, and Q is the Brocard circle, and so the Brocard circle does indeed go through Ω, and similarly through Ω'. ∎

Obviously then, the center of the Brocard circle is the midpoint of KO. Recalling that the center of the first Lemoine circle is also the midpoint of KO, we note in passing that these two circles are concentric.

4. The Brocard triangles

(a) A FEW OBSERVATIONS.

(i) We have seen that the lines $B\Omega$ and $C\Omega$ give the points P and Q on the Brocard circle. Similarly, $A\Omega$ determines a point R on the circle, and the triangle PQR is called the *First Brocard Triangle* of $\triangle ABC$ (Figure 139).

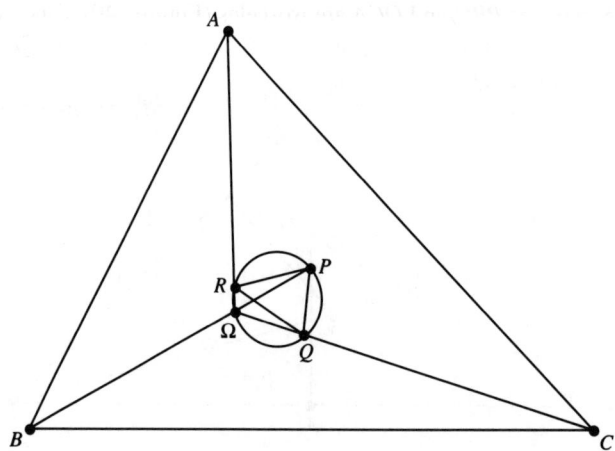

Figure 139

THE BROCARD POINTS

We have also seen that P and Q are points on the perpendicular bisectors of BC and AC. Similarly R is on the perpendicular bisector of AB, and hence the first Brocard triangle is also given by the appropriate three points of intersection of the Brocard circle and the perpendicular bisectors of the sides of a triangle.

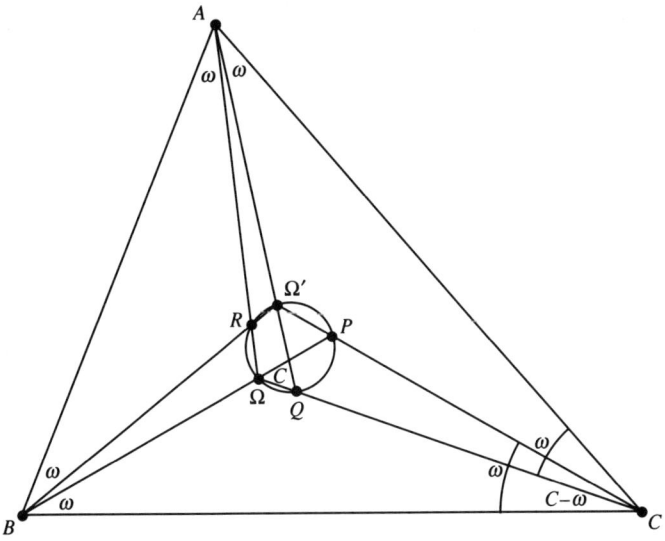

Figure 140

(ii) Now, because P and Q are respectively on the perpendicular bisectors of BC and AC, the triangles PBC and QCA are isosceles (Figure 140). Since Ω lies on BP, the equal base angles in $\triangle PBC$ are ω, and similarly, Ω lies on QC, making $\omega = \angle QCA = \angle QAC$.

Thus, since the definition of the second Brocard point Ω' requires that each of the angles $\angle \Omega'CB$ and $\angle \Omega'AC$ be ω, it follows that Ω' lies on each of PC and AQ. Hence

P is the point of intersection of $B\Omega$ and $C\Omega'$,

and

Q is the point of intersection of $C\Omega$ and $A\Omega'$.

Similarly, R is the point of intersection of $A\Omega$ and $B\Omega'$, and we have the striking result that

> it doesn't matter which Brocard point is used to generate the first Brocard triangle—the same three points P, Q, and R on the Brocard circle are obtained whether you join the vertices to Ω or to Ω'.

(iii) Referring to Figure 140, since $\angle \Omega CA = \omega$, we have $\angle \Omega CB = C - \omega$. Hence, in $\triangle \Omega BC$,

$$\text{exterior angle } P\Omega Q = \omega + (C - \omega) = C.$$

Now, in the Brocard circle, the chord PQ subtends the same angle at Ω and R, and so $\angle R$ in the Brocard triangle is equal to $\angle C$. Note that R lies on the perpendicular bisector of AB, and C is the angle opposite AB in $\triangle ABC$. Similarly, then, the angles P and Q in $\triangle PQR$ are respectively equal to angles A and B in $\triangle ABC$. Thus triangles PQR and ABC are similar. However, since the cyclic directions $A \rightarrow B \rightarrow C$ and $P \rightarrow Q \rightarrow R$ are opposite in the two triangles, we have that

a triangle and its first Brocard triangle are inversely similar (Figure 141).

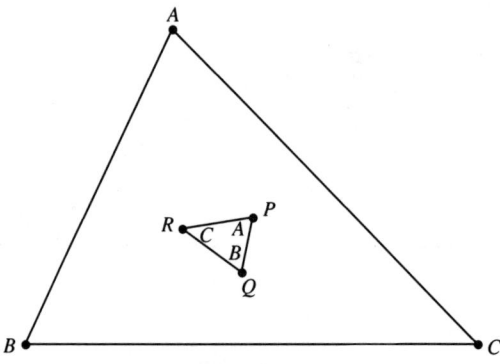

Figure 141

(b) Now let us prove the unexpected property that

a triangle ABC and its first Brocard triangle PQR have the same centroid.

Observe that, with Ω lying on AR, BP, and CQ, and Ω' on AQ, BR, and CP, the Brocard angle ω occurs in the six places shown in Figure 142. Thus the three isosceles triangles RAB, PBC, and QCA have equal base angles and therefore are similar, and we obtain the equal ratios

$$\frac{\text{an arm}}{\text{the base}} = \frac{BP}{a} = \frac{CQ}{b} = \frac{BR}{c}.$$

Thus

(1) $$\frac{BR}{BP} = \frac{c}{a} \quad \text{and} \quad \frac{CQ}{BP} = \frac{b}{a}.$$

THE BROCARD POINTS

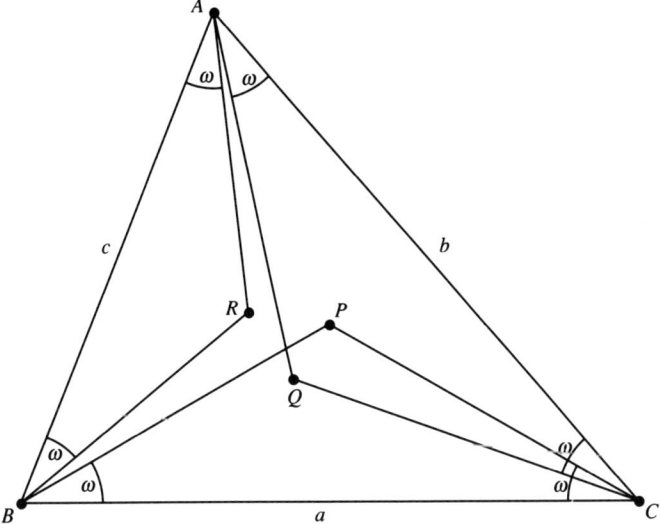

Figure 142

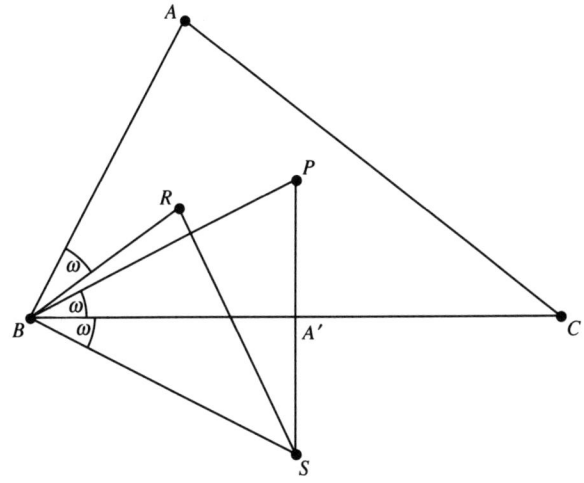

Figure 143

We have seen that PA' is the perpendicular bisector of BC (Figure 143). Now let P be reflected in BC to give S. This makes

$$\angle SBC = \angle PBC = \omega,$$

and hence

$$\angle RBS = \angle B.$$

But $BP = BS$, and so

$$\frac{BR}{BS} = \frac{BR}{BP} = \frac{c}{a} \quad \text{by (1)},$$

and we see that the sides about the equal angles B in triangles ABC and RBS are proportional (Figure 144). Therefore these triangles are similar, and so in $\triangle RBS$ we have $\angle C$ at S and $\angle A$ at R.

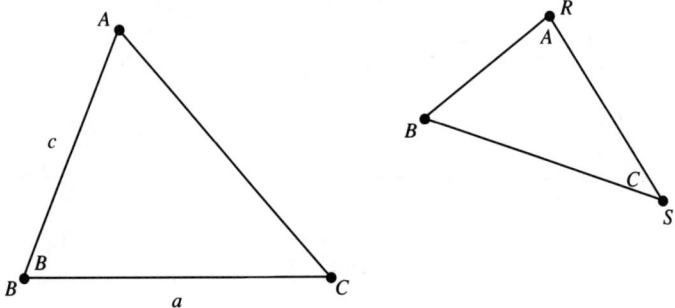

Figure 144

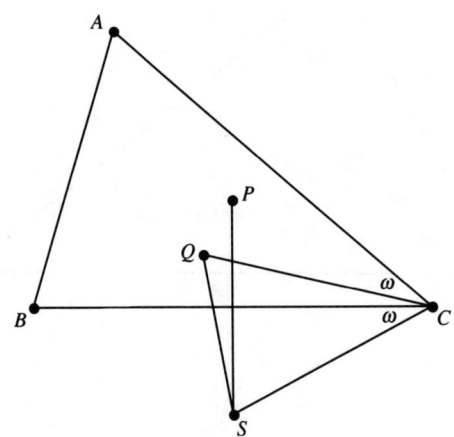

Figure 145

In the same way,

$$\frac{CQ}{CS} = \frac{CQ}{CP} = \frac{CQ}{BP} = \frac{b}{a},$$

THE BROCARD POINTS

and $\triangle QSC$ is also similar to $\triangle ABC$, with $\angle B$ at S and $\angle A$ at Q (Figures 145 and 146). Therefore triangles QSC and RBS are similar. However, since the corresponding sides BS and CS (opposite $\angle A$) are equal, the triangles are actually congruent, and it follows that

$$AR = BR = QS \quad \text{and} \quad AQ = QC = RS \quad \text{(Figure 146)}.$$

Thus, in quadrilateral $ARSQ$, opposite sides are equal.

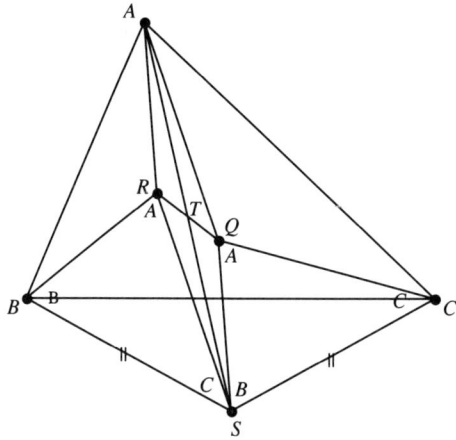

Figure 146

This is not enough to prove that $ARSQ$ is a parallelogram, for it could be self-intersecting. However, as we shall see, AR and QS are also parallel, for each is inclined to AB at the Brocard angle ω.

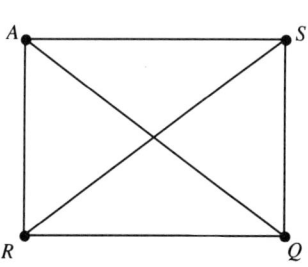 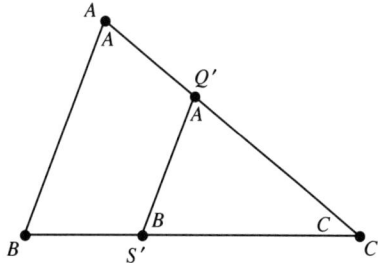

Figure 147

It has already been observed that AR is so inclined to AB. Now since $\angle SCB = \angle QCA = \omega$ (Figure 145), the rotation $C(-\omega)$ takes CQ and CS respectively to lie along CA and CB; that is to say, it takes $\triangle QSC$ to a triangle $Q'S'C$ as shown in Figure 147. Since $\angle Q = \angle A$ in $\triangle QSC$, then $\angle S'Q'C = \angle A$, making $Q'S'$ parallel to AB. Thus in its initial direction, QS must also have been inclined to the direction of AB at the angle ω, and therefore AR and QS are parallel. In parallelogram $ARSQ$, then, the diagonals AS and QR bisect each other at a point T (Figure 146).

Therefore in the Brocard triangle PQR, PT is the median to QR (Figure 148), and in $\triangle APS$, PT is the median to AS. But because the centroid of a triangle trisects each median, triangles which share a median from a common vertex must have the same centroid. Hence $\triangle PQR$ has the same centroid G as $\triangle APS$. But because $PA' = A'S$, one of the medians of $\triangle APS$ is AA', which is also a median of $\triangle ABC$, and therefore G is also the centroid of $\triangle ABC$. Thus the first Brocard triangle, the given triangle ABC, and $\triangle APS$ all have the same centroid.

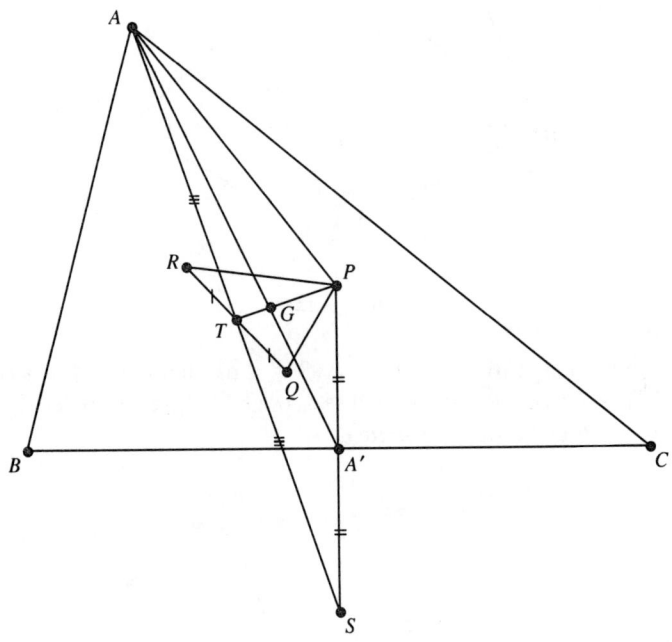

Figure 148

A REMARKABLE CONCURRENCE. As above, let the first Brocard triangle of $\triangle ABC$ be PQR, where P is the vertex that lies on the perpendicular bisector of BC, Q the vertex on the perpendicular bisector of AC, and R on the perpendicular bisector of AB. Let perpendiculars be drawn from the midpoints U, V, W of the

THE BROCARD POINTS

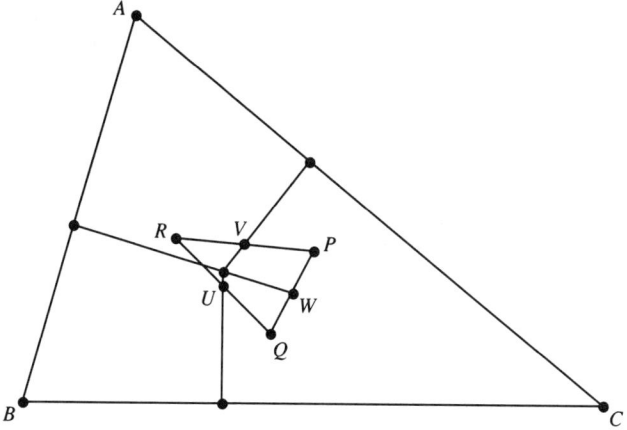

Figure 149

sides of $\triangle PQR$ (opposite P, Q, R, respectively, Figure 149) to the sides BC, AC, and AB of $\triangle ABC$ (opposite A, B, C, respectively). Then

these three perpendiculars are concurrent.

And guess where they meet! Who would imagine that

they meet at the center N of the nine-point circle of $\triangle ABC$.

The proof is surprisingly easy. We know that N bisects the segment HO on the Euler line, and that the centroid G trisects it (Figure 150). Thus the dilatation $G(-\frac{1}{2})$ takes O to N. Since G is also the centroid of $\triangle PQR$ (proved in the

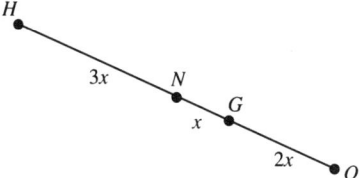

Figure 150

preceding section), $G(-\frac{1}{2})$ takes P into the midpoint U of the opposite side QR (Figure 151). Thus the dilatation takes PO to UN. But a dilatation doesn't alter the direction of a line; and since PO lies along the perpendicular bisector PA' of BC, so the perpendicular from U to BC is the line UN. Similarly the other perpendiculars are VN and WN, establishing the concurrence. ∎

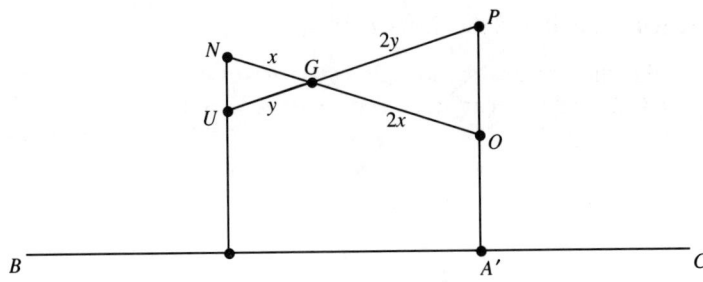

Figure 151

(c) THE SECOND BROCARD TRIANGLE.

We have seen that the first Brocard triangle is determined on the Brocard circle by the lines $A\Omega$, $B\Omega$, and $C\Omega$. If the symmedian point K is used instead of Ω, then it is the *second Brocard triangle* that is determined on the Brocard circle (Figure 152).

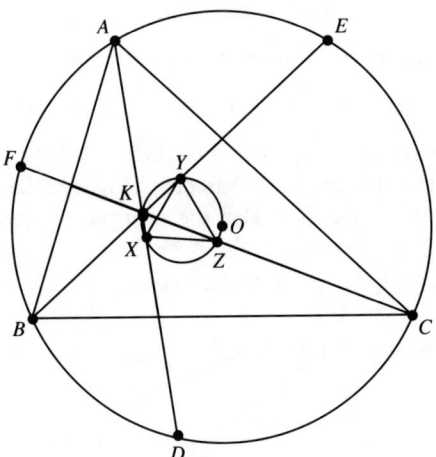

Figure 152

Let the lines AK, BK, and CK be extended to meet the circumcircle of $\triangle ABC$ at D, E, and F; then the vertices of the second Brocard triangle are the *midpoints* of the chords AD, BE, and CF.

The proof is immediate. Because KO is a diameter of the Brocard circle, each of the angles KXO, KYO, and KZO is a right angle, and the perpendicular from the center of a circle to a chord bisects the chord.

5. The Steiner point and the Tarry point

(a) Recall that the first Brocard triangle PQR is inversely similar to $\triangle ABC$, and the angles A, B, and C occur at the vertices P, Q, and R respectively. Now let a line through A be drawn parallel to the side of $\triangle PQR$ opposite its angle A, i.e. parallel to QR, and let similar lines be drawn through B and C. Then these lines are concurrent at a point called the *Steiner point of S of $\triangle ABC$*. Can you guess the whereabouts of S? The answer is not so startling this time: *S lies on the circumcircle of $\triangle ABC$*.

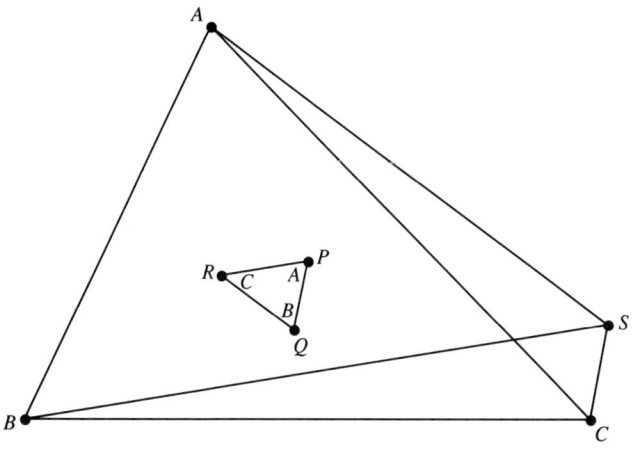

Figure 153

Let the lines constructed through A and B meet at W (Figure 154(a)). Clearly these lines meet at the same angle as their corresponding parallels RQ and RP,

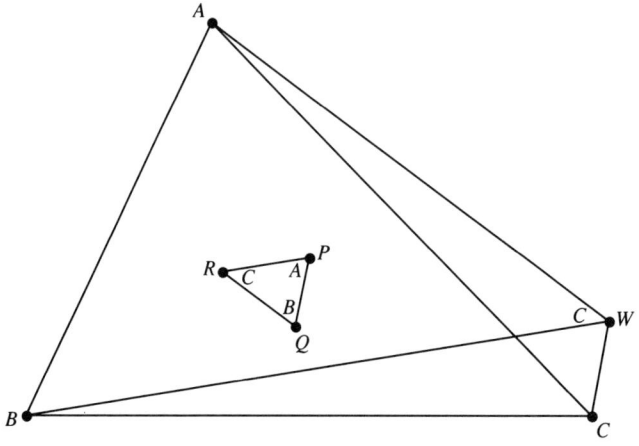

Figure 154(a)

i.e. ∠C. Thus AB subtends the same angle at C and W, making ABCW cyclic. Hence BC subtends the same angle at A and W, showing that CW is inclined to BW at ∠A. Now the angle between RP and PQ in △PQR is ∠A, and since RP is parallel to BW, it follows that CW is the parallel to PQ at C, establishing the desired concurrence at W on the circumcircle of △ABC. ∎

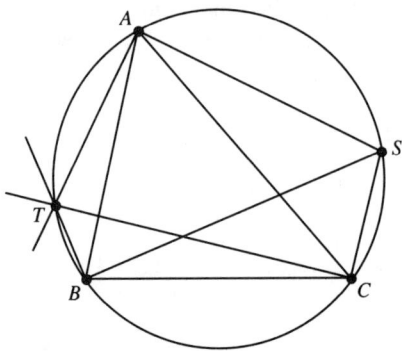

Figure 154(b)

If *perpendiculars* are drawn through A, B and C instead of parallels, each line will go through the point T diametrically opposite S on the circumcircle (Figure 154(b)). This point T is called the *Tarry point* of △ABC.

We note in passing that

the Steiner point of a triangle is the center of mass of the system obtained by suspending at each vertex a mass equal to the magnitude of the exterior angle at that vertex.

(b) A SURPRISE.

Now consider the first Brocard triangle $P'Q'R'$ of △ABC's first Brocard triangle PQR (Figure 155). Since △$P'Q'R'$ is inversely similar to △PQR, which is itself inversely similar to △ABC, it follows that △$P'Q'R'$ is *directly* similar to △ABC.

Thus △PQR with △$P'Q'R'$ inside it is a smaller copy of △ABC with △PQR inside it; however, since triangles PQR and ABC are inversely similar, the copy is turned over on the page (Figure 155). Now the angles A, B, and C occur at the corresponding vertices of △PQR, in particular ∠B occurs at Q, and the angle θ at which PQ meets BC occurs again as the angle between $P'Q'$ and QR. Thus we can deduce that $P'Q'$ is parallel to AB as follows.

Let BC be rotated about L through the angle θ to lie along LQP; then let LQP be rotated about Q through ∠B to lie along $QL'R$; finally, rotate $QL'R$ about L' through the angle $-\theta$ to the direction of $L'Q'P'$. These rotations result in a net change of direction for BC of $\theta + \angle B - \theta = \angle B$, revealing that the final position $P'Q'$ is inclined to BC at the ∠B, as is AB. ∎

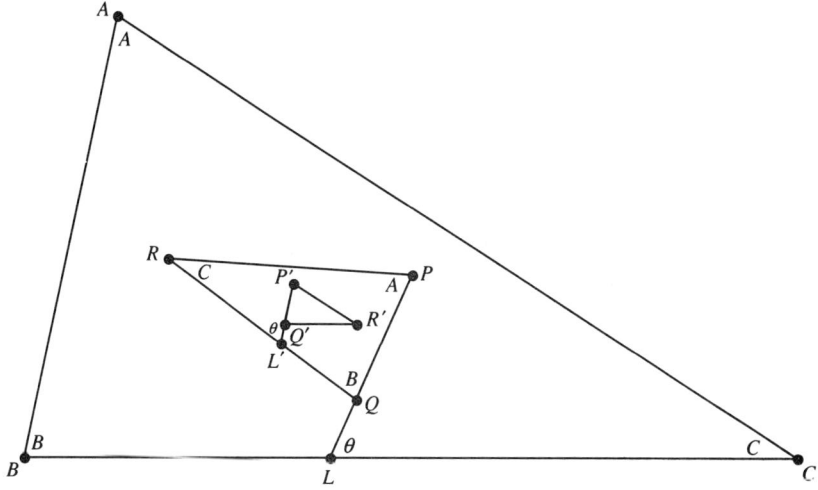

Figure 155

Similarly $Q'R'$ is parallel to BC, and $P'R'$ to AC, and so triangles $P'Q'R'$ and ABC are not only directly similar, but their corresponding sides are parallel.

Now the angles A, B, C of $\triangle ABC$ occur again at the corresponding vertices P, Q, R in $\triangle PQR$ and at P', Q', R' in $\triangle P'Q'R'$. Accordingly, the Steiner point of $\triangle PQR$ is the point of intersection of lines through P, Q, and R which are parallel respectively to the sides $Q'R'$, $P'R'$, and $P'Q'$ of its first Brocard triangle $P'Q'R'$. Because the corresponding sides of triangles ABC and $P'Q'R'$ are parallel, these lines through P, Q and R are therefore respectively parallel to BC, AC, and AB. But, by an earlier result (pp. 109–110, Figure 138), such parallels are given by the lines PK, QK, and RK to the symmedian point K (Figure 156), and we have the beautiful result that the symmedian point K of $\triangle ABC$ is the Steiner point of $\triangle PQR$, that is,

> the symmedian point of a triangle is the Steiner point of its first Brocard triangle (and O is its Tarry point).

We note in passing that the Simson lines of the Steiner and Tarry points of a triangle ABC are respectively parallel and perpendicular to the Brocard diameter KO.

6. A property relating K, G, Ω, Ω'

The symmedian point K, the centroid G, and the Brocard points Ω and Ω' are certainly four of the most noteworthy points of a triangle. Let us conclude this excursion into Brocard geometry with a result that connects these points in a most unexpected and appealing way.

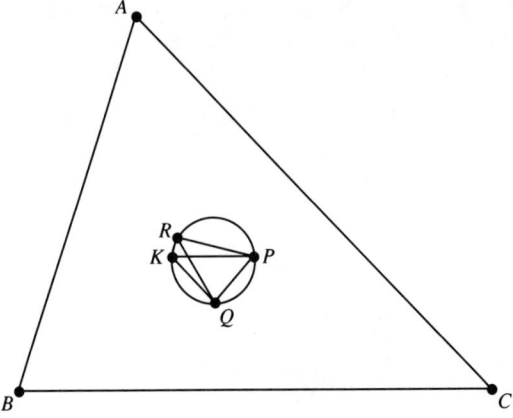

Figure 156

A line from a vertex to a Brocard point is called a *Brocard ray*; thus there are two Brocard rays from each vertex. Now, it is a remarkable fact that

> the symmedian from one vertex of a triangle, the median from another, and the appropriate Brocard ray from the third vertex are concurrent.

Let's establish the property for the symmedian from B, the median from C and the Brocard ray $A\Omega$ (Figure 157).

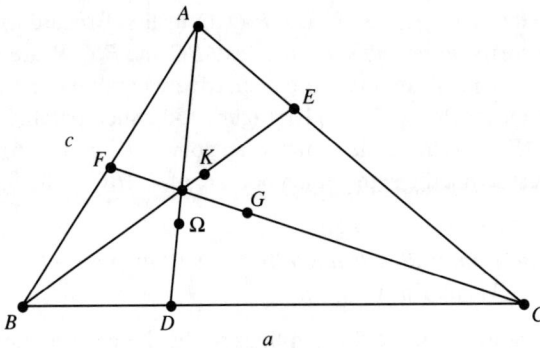

Figure 157

If these lines divide the sides of $\triangle ABC$ into ratios whose product is 1, the conclusion would follow by Ceva's theorem. Let us show, then, that

$$\frac{AF}{FB} \cdot \frac{BD}{DC} \cdot \frac{CE}{EA} = 1.$$

Now the median CF bisects AB, and because a symmedian divides a side into segments proportional to the *squares* of adjacent sides, we have

$$\frac{CE}{EA} = \frac{a^2}{c^2}.$$

Since $AF = FB$, it remains, then, to show only that

$$\frac{BD}{DC} = \frac{c^2}{a^2}.$$

Now each Brocard ray is inclined to a side of $\triangle ABC$ at the Brocard angle ω. Therefore let the angles and segments be labelled as in Figure 158, where the appearance of angles A and B at Ω needs to be justified.

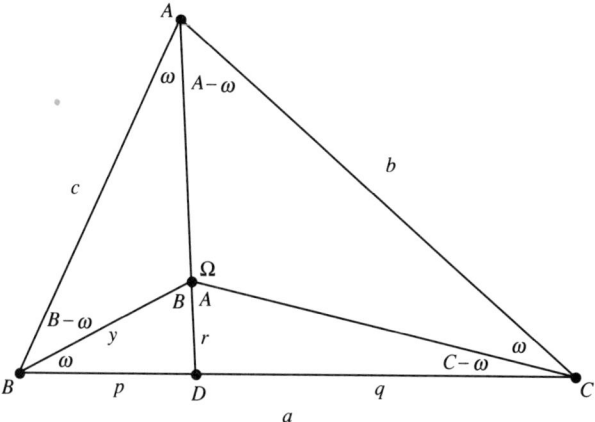

Figure 158

Since triangles $B\Omega D$ and BAD have a common angle at D, and each has a second angle equal to ω, the triangles are similar, and therefore their third angles are equal, $\angle B\Omega D = \angle B$, and we have the proportion

(2) $$\frac{y}{r} = \frac{c}{p}.$$

Now, since $\angle BA\Omega = \omega$, then $\angle \Omega AC = A - \omega$, and with $\angle AC\Omega = \omega$, the exterior angle $D\Omega C$ of $\triangle \Omega AC$ is $(A - \omega) + \omega = A$. Hence angle $B\Omega C = A + B$. Applying the law of sines to triangles $B\Omega C$ and $D\Omega C$, we have

$$\frac{y}{\sin(C - \omega)} = \frac{a}{\sin(A + B)} \left(= \frac{a}{\sin C} \right),$$

and
$$\frac{r}{\sin(C-\omega)} = \frac{q}{\sin A}.$$

Dividing these equations we get
$$\frac{y}{r} = \frac{a \cdot \sin A}{q \cdot \sin C} = \frac{a}{q} \cdot \frac{a}{c} = \frac{a^2}{qc}.$$

From the proportion (2), then
$$\frac{c}{p} = \frac{y}{r} = \frac{a^2}{qc},$$

from which we obtain the desired
$$\frac{p}{q} = \frac{c^2}{a^2}. \quad \blacksquare$$

CHAPTER ELEVEN

The Orthopole

1. Let perpendiculars *AL*, *BM*, and *CN* be drawn to a straight line *m* from the vertices of triangle *ABC* (Figure 159). From *L*, the foot of the perpendicular from *A*, draw *LX* perpendicular to *BC*, the side of △*ABC* opposite *A*. Similarly, draw *MY* perpendicular to *AC* and *NZ* perpendicular to *AB*. Then,

the perpendiculars from L, M, and N concur at a point called the orthopole of m and △ABC.

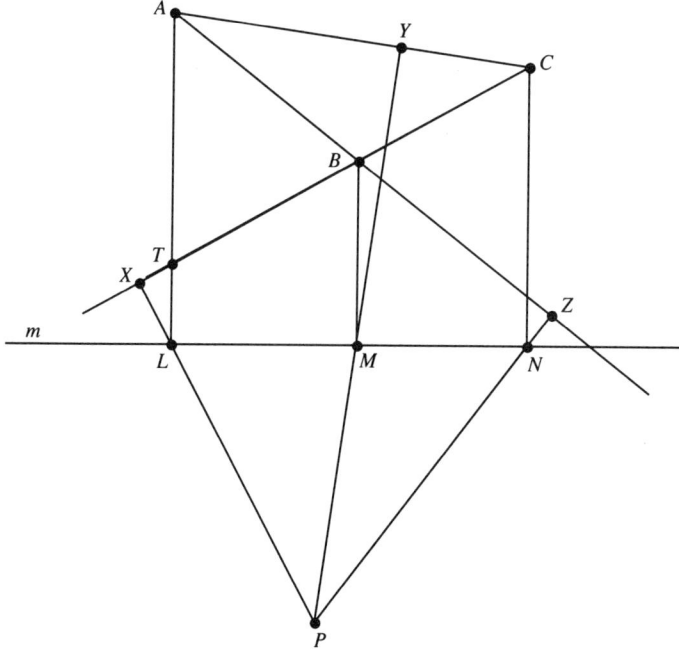

Figure 159

To begin the proof, leave out *NZ*, and denote the intersection of *LX* and *MY* by *Q* (Figure 160); also, let *AL* meet *BC* at *T*. Then the sides of △*TAC* are perpendicular to the sides of $\triangle LMQ$, making the triangles similar. In these triangles,

$$\angle C = \angle Q \quad (= \theta), \quad \angle T = \angle L \quad (= \phi),$$

and therefore

(1) $$\frac{AT}{TC} = \frac{LM}{LQ}.$$

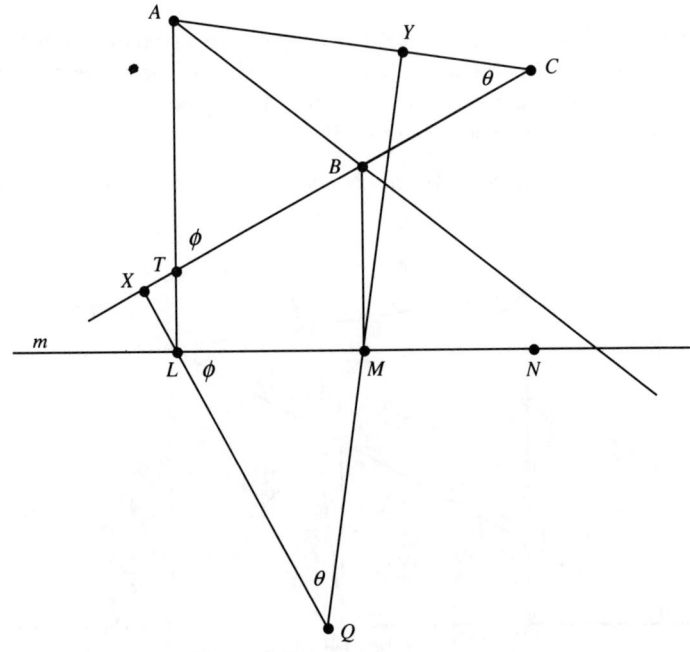

Figure 160

Now let's start again, this time leaving out the perpendicular MY, and denoting the intersection of LX and NZ by R (Figure 161). Then triangles TAB and LNR have perpendicular sides and are similar, and

(2) $$\frac{TB}{AT} = \frac{LR}{LN}.$$

But the parallels AL, BM, and CN (see Figure 159) give a third proportion:

(3) $$\frac{TC}{TB} = \frac{LN}{LM}.$$

THE ORTHOPOLE

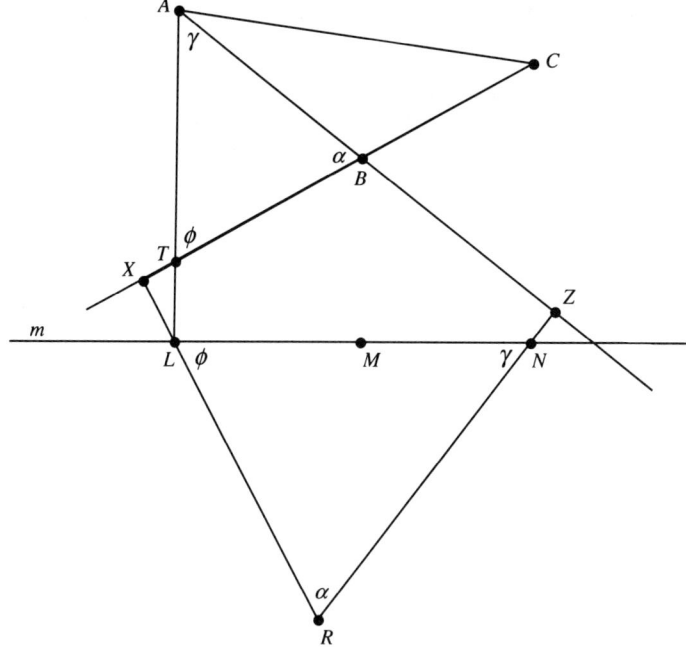

Figure 161

Multiplying these three proportions, we get

$$\frac{LM}{LQ} \cdot \frac{LR}{LN} \cdot \frac{LN}{LM} = \frac{AT}{TC} \cdot \frac{TB}{AT} \cdot \frac{TC}{TB},$$

$$\frac{LR}{LQ} = 1,$$

giving

$$LR = LQ.$$

Since both Q and R are on LX, Q and R are the same point thus establishing the orthopole. ∎

2. Now the orthopole has two properties I hope you will enjoy; the first is a delightful surprise and the second is a real gem:

(a) the orthopole of a line through the circumcenter lies on the *nine-point circle* of the triangle.
(b) if a line crosses the circumcircle of a triangle, the Simson lines of the points of intersection meet at the orthopole of the line.

In proving (b), we will get part (a) as an easy bonus. In order to establish (b), however, we will need several preliminary results which have their own appeal.

(i) To begin, let's investigate what happens to the orthopole when the line m is moved parallel to itself. Let P_1 be the orthopole of m_1 relative to $\triangle ABC$ (Figure 162; any two of the defining perpendiculars LX, MY, and NZ suffice to determine an orthopole). Now let m_1 be moved to a parallel position m_2. The perpendiculars BM_2 and CN_2 lie on the same lines as BM_1 and CN_1, respectively, and the orthopole P_2 of m_2 is the intersection of the perpendiculars from M_2 and N_2 to sides AC and AB. Instead of drawing these perpendiculars, let P_1Q be drawn equal and parallel to M_1M_2. Then $P_1QM_2M_1$ is a parallelogram, and since M_1P_1 is perpendicular to AC, M_2Q is also perpendicular to AC. Similarly, N_2Q is perpendicular to AB, making Q the orthopole of m_2.

Thus the orthopole P_1 moves along a line *perpendicular* to m_1 a distance equal to the displacement from m_1 to m_2. As m_1 ranges over all parallel positions, then, the orthopole traces out the entire perpendicular line.

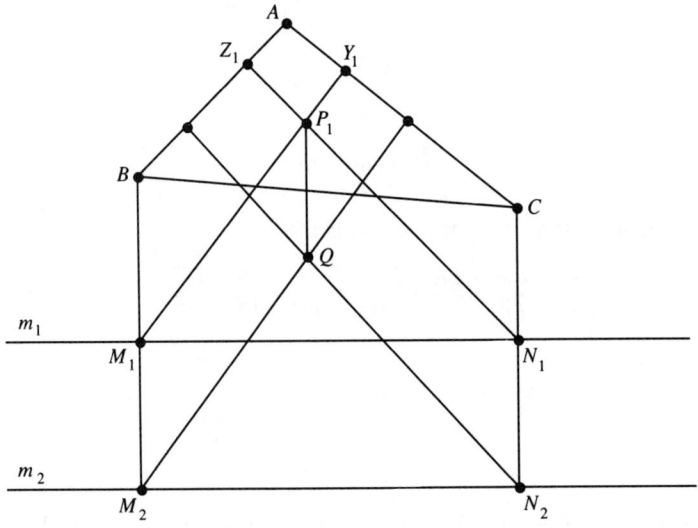

Figure 162

(ii) If ℓ is the Simson line of a point P, then P is called the *pole* of ℓ. Next, let's consider a simple construction for the pole of a Simson line whose direction d is prescribed.

Through any vertex of the triangle, say A, draw the chord AD in the given direction d (Figure 163(a)). From D, draw the chord DP perpendicular to the side opposite A, in this case BC, to get the pole P.

THE ORTHOPOLE

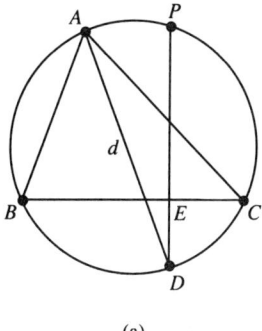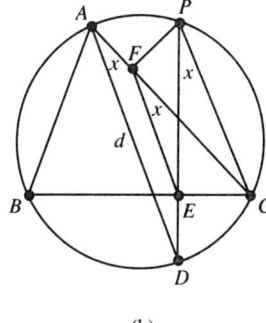

(a)   (b)

Figure 163

To prove that this is the pole of a Simpson line with given direction d, draw the perpendicular PF to AC (Figure 163(b)). If DP crosses BC at E, then EF is the Simson line of the point P. Now the right angles at E and F imply that $PFEC$ is cyclic, and the angles marked x at F and P are equal. But in the given circle, the angles x on arc DC at A and P are also equal, and we have equal corresponding angles at A and F implying $AD \parallel FE$. Thus the Simson line EF has the prescribed direction d. ∎

Conversely, starting with a specified point P, the *direction* of its Simson line is found simply by reversing this construction. Since different poles P determine different chords PD, and hence different directions AD, we see there is *only one* Simson line in a specified direction.

(iii) Recall that we showed in Chapter 5, Section 4 that *the Simson line of P passes through the midpoint of the segment PH to the orthocenter H of the triangle.* We also noted there that the midpoint of PH lies on the nine-point circle of the triangle.

(iv) Now let m be a line through the circumcenter O of $\triangle ABC$ and let P be the pole of the Simson line *perpendicular* to m (Figure 164; thus P is determined by drawing AD perpendicular to m and then DP perpendicular to BC). We will show that *the orthopole of m is the midpoint K of PH*, which, as we have just recalled, lies on the nine-point circle. This will establish property (a).

To prove this, we will show that each of the perpendiculars LX, MY, and NZ in the definition of "orthopole" goes through K. If AD crosses m at L, only the typical case of LX need be considered (Figure 164).

Since OL is the perpendicular to chord AD from the center O, L is the midpoint of AD. Now AHT, LX and PED are all perpendicular to BC, and so LX, in bisecting AD, is the midline of the strip determined by the other two. Thus LX bisects every segment across this strip, in particular PH, and hence goes through K.

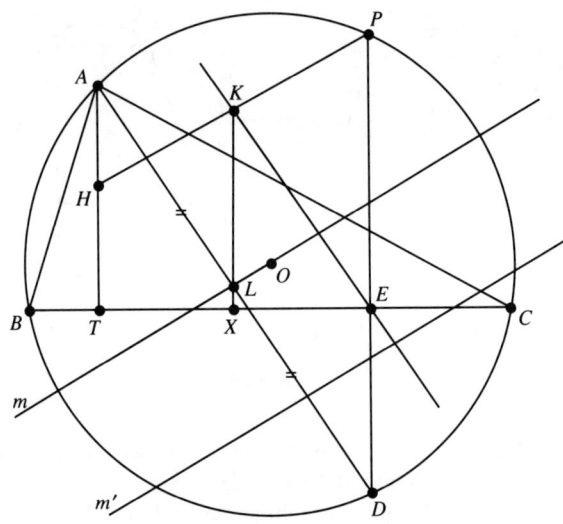

Figure 164

(v) Since a Simson line bisects the segment *PH*, the orthopole *K* also lies on the Simson line of *P*, and in Figure 164, *KE* is in fact the Simson line of *P*.

Now let *m* move parallel to itself to any position *m'* (Figure 164). As shown earlier, this causes the orthopole *K* to slide along the line perpendicular to *m* and *m'*, in this case the perpendicular Simson line *KE*. Since variations in *m* and *m'* result in *m'* ranging over all lines in the plane, we have the important result that

the orthopole of a line lies on the Simson line which is perpendicular to it.

We have seen that there is only one Simson line in a given direction. Hence this result is equivalent to the assertion that, *among all the perpendiculars to a line m, the Simson line is the one through the orthopole of m.*

Finally, we are ready to proceed with the ingenious argument to establish property (b).

(vi) Let *m* intersect the circle at *Q* and *R* and let *P* be the pole of the Simson line perpendicular to *m* (Figure 165). The foot *E* of the perpendicular from *P* to *BC* lies on the Simson line of *P*, and so the perpendicular *EZ* to *m* is in fact the Simson line of *P*. Since the perpendicular Simson line and also the defining perpendicular *LX* both pass through the orthopole *K* of *m*, it follows that *LX* meets *EZ* at *K*.

Now, one point on the Simson line of *Q* is the foot *S* of the perpendicular to *BC*. Let *QS* be extended in both directions to give *T* on the circle and to meet the perpendicular from *D* to *QS* at *V*. Since our goal is to show that the Simson line of *Q* goes through the orthopole *K*, we need to show that *SK* is actually the

THE ORTHOPOLE

Simson line of Q. Because the Simson line goes through S, it remains only to show that SK lies in the **right direction** to be the Simson line of Q.

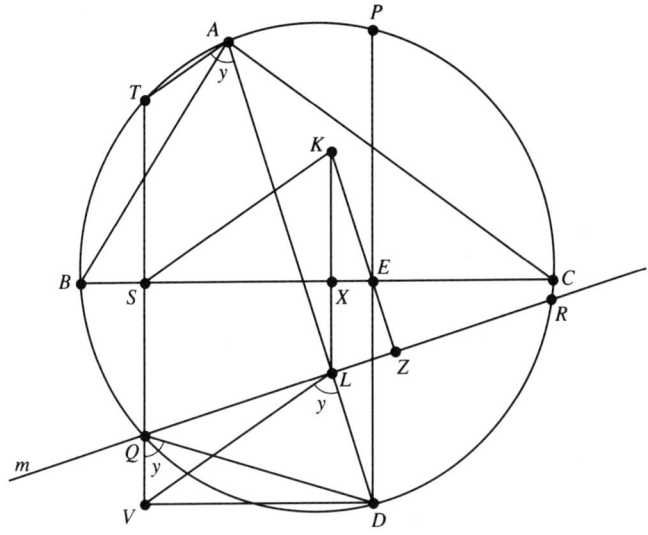

Figure 165

Observe that the right angles at V and L make $QVDL$ cyclic and give

$$\angle VQD = \angle VLD \quad (= y).$$

But $ATQD$ is inscribed in the given circle, and so the exterior angle VQD is equal to the interior angle TAD at A. Thus the equal corresponding angles y at A and L show that TA and VL are parallel.

Now the opposite sides of $KLDE$ are clearly parallel, making it a parallelogram. Hence KL is equal and parallel to DE. But, in rectangle $EDVS$, DE is equal and parallel to SV, and so KL and SV are equal and parallel. Therefore $SVLK$ is a parallelogram and SK is parallel to VL. Since VL is parallel to TA, we have the crucial fact that SK is parallel to TA.

Now we ask "How would you construct the Simson line in the direction of TA?". According to the method described above, we could begin by drawing the chord from A in the given direction—that would give AT itself. Then we would continue by drawing the perpendicular from T to the side BC opposite A—that would give TSQ—and the pole of the desired Simson line would be obtained at Q. That is to say, the Simson line of Q lies in the direction of TA, and since this is also the direction of SK, we conclude that SK *does* have the proper direction and is indeed the Simson line of Q. Similarly, the Simson line of R goes through K and our argument is complete. ∎

3. The Rigby Point

Let's close this essay with some exciting further developments due to John Rigby.

(a) Let QR be a chord of the circumcircle of $\triangle ABC$ and let P be the pole of the Simson line S_P which is *perpendicular* to QR (given by ML in Figure 166). Since the orthopole of a line lies on the Simson line perpendicular to the line, the orthopole X of QR must lie on S_P. We have just proved that the Simson lines S_Q and S_R of Q and R meet at the orthopole X of QR. Thus the Simson lines of P, Q, and R all meet at X.

But this is just the beginning. You might be surprised to learn that X is also the orthopole of PQ and PR and that, interchanging the roles of triangles ABC and PQR, the three Simson lines of A, B, and C, with respect to $\triangle PQR$, all go through this same point X, which also turns out to be the orthopole, relative to $\triangle PQR$, of each of the three sides AB, BC, and AC of $\triangle ABC$.

This "Rigby Point" X is thus the orthopole of the six sides of triangles ABC and PQR and lies on the six Simson lines of the vertices!

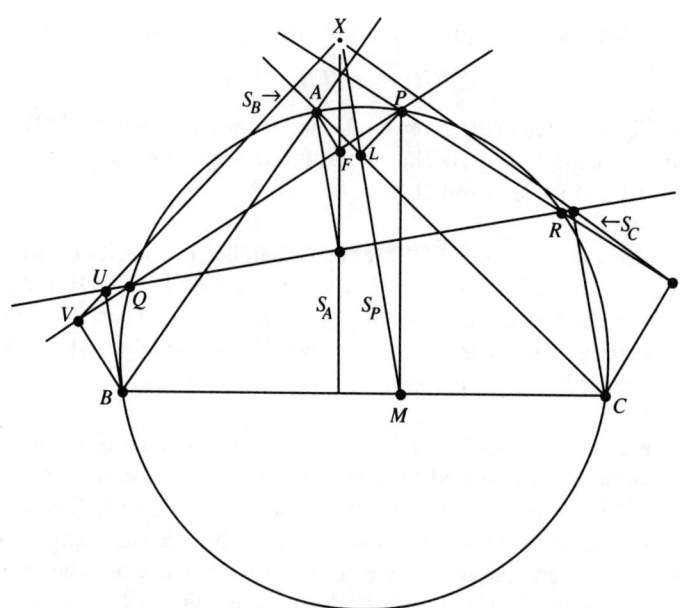

Figure 166

THE ORTHOPOLE

The key to all this is the fact that, by insisting that P be the pole of the Simson line which is perpendicular to QR, we fall heir to five additional such perpendicularities. Each of the six Simson lines is perpendicular to the side of $\triangle ABC$ or $\triangle PQR$ which is *opposite* its pole; thus

relative to $\triangle ABC$ $\begin{cases} \text{the Simson line of } P \text{ is perpendicular to } QR, \\ \text{the Simson line of } Q \text{ is perpendicular to } PR, \\ \text{the Simson line of } R \text{ is perpendicular to } PQ, \end{cases}$

relative to $\triangle PQR$ $\begin{cases} \text{the Simson line of } A \text{ is perpendicular to } BC, \\ \text{the Simson line of } B \text{ is perpendicular to } AC, \\ \text{the Simson line of } C \text{ is perpendicular to } AB. \end{cases}$

(b) Let's begin with the easy proof that the Simson line S_A of A, that is, FG in Figure 167, is perpendicular to BC. Clearly AG and LM are parallel since each is perpendicular to QR, and if we show that FG is inclined to AG at the same angle that PM is inclined to LM, i.e. $x = w$, FG would be parallel to PM. Since PM is perpendicular to BC, then the desired conclusion would follow.

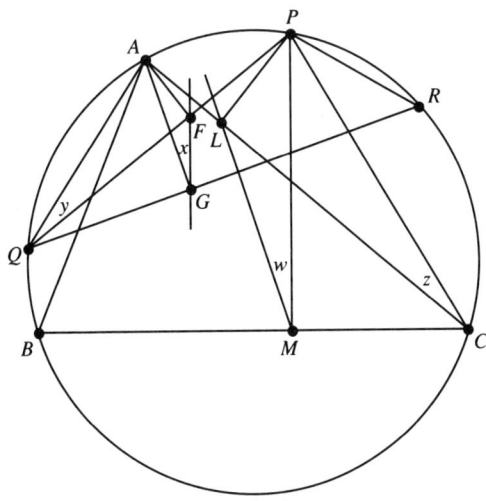

Figure 167

Now AQ subtends right angles at F and G, making $AQGF$ cyclic, and giving equal angles x and y on chord AF. But, in the given circle, y and z are equal angles on chord AP. Finally, PC subtends right angles at L and M, making $PLMC$ cyclic, and giving equal angles z and w on chord PL. Thus $x = y = z = w$, as desired. ∎

Similarly, the Simson lines S_B and S_C are perpendicular to AC and AB, respectively.

(c) Given QR, P is defined so as to make S_P perpendicular to QR. It is somewhat involved, however, to prove that S_Q is perpendicular to PR. In Figure 168, if we could show that S_Q is inclined to S_P at the same angle θ that PR is inclined to QR, the conclusion would follow.

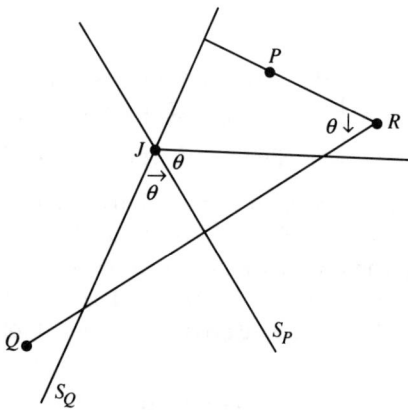

Figure 168

Now, we are only justified in claiming the perpendicularity of S_Q and PR if their directions are obtained from the known perpendiculars by rotations through the same angle in the same direction. Our argument would fail if S_Q were to lie on the wrong side of S_P. (Strictly speaking, we should have addressed this issue every time this argument was used.) It is *directed angles* that are involved and, in Figure 168, the result we need is that the positive angle from S_Q to S_P is the same as the positive angle from PR and QR (which we will write as $\angle PRQ$, in contrast to the negatively directed $\angle QRP$). This is not difficult to show, but it does require a slight digression.

In Figure 169, observe that, relative to $\triangle ABC$, the Simson line S_A of the vertex A itself is just the altitude from A, and therefore its direction is given by the perpendicular PM that arises in the construction of S_P. Thus, as A moves clockwise around the circumference to a point P, the Simson line of the moving point spins counterclockwise from the direction of PM to that of LM.

Now the right angles at L and M make $PLMC$ cyclic, giving equal angles θ at M an C on chord LP. That is to say, to go *from* the "standard direction" of S_A *to* the direction of S_P requires a rotation in a *positive* sense through an angle equal to the angle that is subtended at the circumference of the circumcircle of $\triangle ABC$ by the traversed arc AP. The magnitude of the angle between two Simson lines S_P and S_Q, then, is given by the difference between the angles θ and ϕ

THE ORTHOPOLE

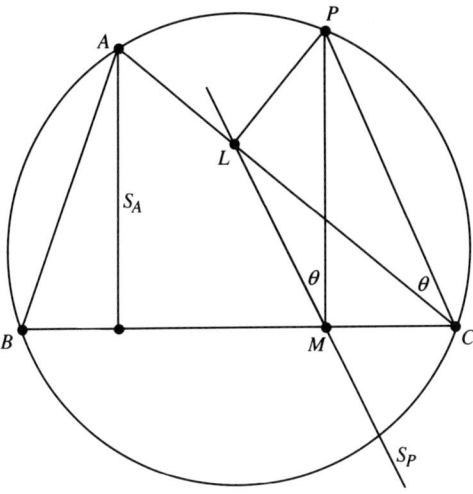

Figure 169

subtended by the arcs AP and AQ, which is just the angle δ subtended by the arc PQ (Figure 170). It is also clear that if P is the interior point and Q the endpoint of the clockwise arc from A, the directed angle from S_P to S_Q is $+\delta$ rather than $-\delta$.

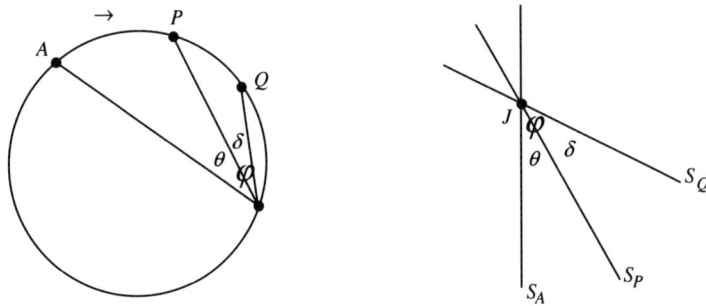

Figure 170

Returning to the proof, the directed angle from S_P to S_Q is $+\delta$, where δ is the angle subtended by the clockwise arc PQ (Figure 171(a)). Recall that our goal is to show that, in Figure 168, the directed angle θ from S_Q to S_P is the same as the directed angle from PR to RQ, i.e. $\angle PRQ$. Since the angle from S_P to $S_Q = +\delta$, the desired angle θ, from S_Q to S_P, is $-\delta$. Thus θ is also given by the equivalent positive angle $180° - \delta$ (Figure 171(b)) which, from Figure 171(a), is also seen to be the directed angle PRQ, as desired. This completes the proof that the Simson line S_Q is perpendicular to PR. Similarly, the Simson line of R is perpendicular to PQ. ∎

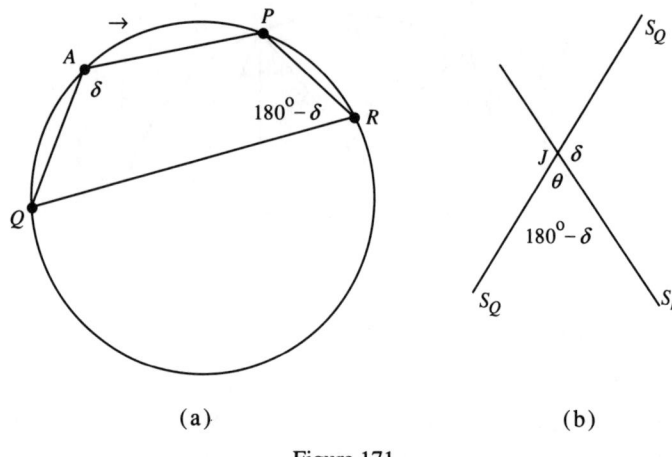

Figure 171

(d) The promised results now follow in rapid succession.

The Simson line S_B, for example, is the line through the feet U and V of the perpendiculars from B to QR and PQ (Figure 166). But, because S_B is perpendicular to AC, it is simply the perpendicular from U to AC.

Now the Rigby point X has been initially defined as the orthopole of QR relative to $\triangle ABC$. In the construction of this orthopole, then, a perpendicular from B to QR would be drawn, thus obtaining the point U, and from U the perpendicular to AC would be drawn as one of the three perpendicular lines that determine X. But this perpendicular is the Simson line S_B, and therefore the Simson line of B does indeed go through X; similarly, so do the Simson lines of A and C. ■

We proved that the orthopole of a chord QR is the point of intersection of the Simson lines of Q and R (see Section 2 of this chapter). Similarly then, the orthopole of the line BC lies on the Simson lines S_B and S_C (all relative to $\triangle PQR$), which we have just observed both pass through X. Hence X is the orthopole of BC, and similarly of AB and AC. ■

Finally, consider the construction of the orthopole of PQ. In Figure 166, perpendiculars from A and B to PQ give the feet F and V, from which perpendiculars would then be drawn respectively to BC and AC. But these perpendiculars are just the Simson lines S_A and S_B, which are known to meet at the Rigby point X. Similarly, S_C goes through X.

In conclusion, here is a pretty exercise.

Exercise

Prove that the Rigby point X is the midpoint of the segment HK which joins the orthocenters of triangles ABC and PQR.

CHAPTER TWELVE

On Cevians

1. Ceva's Theorem

As noted in Chapter 1, a segment which goes from a vertex of a triangle to the opposite side is called a *cevian*, in honor of the 17th century Italian engineer Giovanni Ceva (1647–1734). In 1678, Ceva published a necessary and sufficient condition for the *concurrence* of three cevians, *one from each vertex*. Let the cevians join vertices A, B, C, respectively, to points D, E, F in the opposite sides. Ceva's theorem states that AD, BE, CF concur if the product of the ratios into which the sides are divided by D, E, and F is 1, and conversely; in Figure 172, AD, BE, and CF are concurrent if and only if

$$\frac{AF}{FB} \cdot \frac{BD}{DC} \cdot \frac{CE}{EA} = \frac{a}{b} \cdot \frac{c}{d} \cdot \frac{e}{f} = 1,$$

i.e. iff $ace = bdf$.

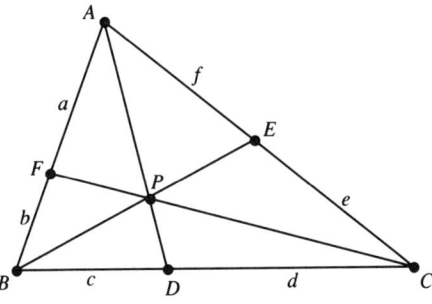

Figure 172

The theorem is easily proved in many books by various methods. I am very fond of the following approach.

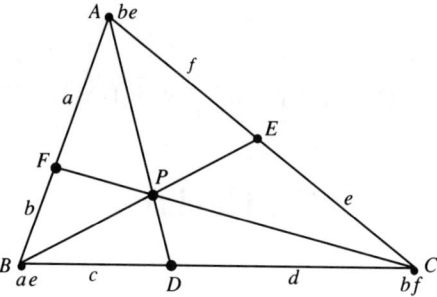

Figure 173

Suspend masses be, ae, and bf at A, B, and C (Figure 173). Since $a(be) = b(ae)$, the moments about F of the masses at A and B are equal, implying that the system is equivalent to masses $(be + ae)$ at F and bf at C. The center of gravity G of the system, then, must lie on CF. Similarly, G lies on BE. The question is whether G lies on AD.

(a) SUPPOSE AD, BE, AND CF CONCUR AT P. Lying on CF and BE, G must be their point of intersection P. If the center of gravity of the masses at B and C is the point X on BC, then the whole system is equivalent to a pair of masses at X and A, and therefore G must occur on the segment AX. That is to say, since $G = P$, X must be in line with A and P, making $X = D$, and the equal moments about D of the masses at B and C give the required $aec = dbf$.

(b) SUPPOSE $aec = dbf$. We have seen that, in any case, the center of gravity G lies on each of CF and BE. Now the equation $aec = dbf$ implies that D is the center of gravity of the masses at B and C, and so G must also lie on AD, establishing the concurrence of AD, BE, and CF. ∎

2. Now let's establish the following two engaging properties of concurrent cevians that are the subject of Problem E1043 in the American Mathematical Monthly, 1958 (page 421):

If AD, BE, and CF are cevians of $\triangle ABC$ through an arbitrary point P inside it, prove that the ratios into which P divides the cevians, namely

$$\frac{AP}{PD}, \frac{BP}{PE}, \frac{CP}{PF},$$

have a sum which is at least 6 and a product which is at least 8 (Figure 173).

(This was proposed by O. J. Ramler, Catholic University, Washington D.C., and solved by Charles W. Trigg, Los Angeles City College.)

ON CEVIANS

This problem can be solved nicely with Ceva's theorem, but the following solution, based essentially on the arithmetic mean-geometric mean inequality, is most incisive and attractive (see the bible of geometric inequalities—Geometric Inequalities, by O. Bottema et al—pages 114 and 115).

Let the areas of the triangles into which $\triangle ABC$ is partitioned by the segments AP, BP, and CP be p, q, and r (Figure 174(a)), and set the areas of the subtriangles BPD and DPC equal to x and y, respectively (Figure 174(b)), so that $x + y = p$.

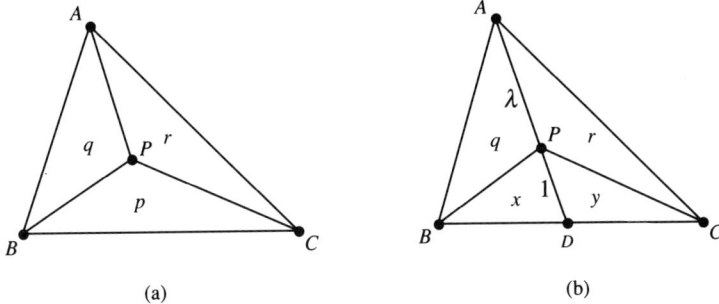

Figure 174

Also, set

$$\frac{AP}{PD} = \lambda, \qquad \text{i.e. } AP = \lambda PD.$$

Then the areas q and x of triangles with the same altitude are proportional to their bases. Hence

$$q = \lambda x \quad \text{and} \quad r = \lambda y,$$

giving

$$q + r = \lambda(x + y) = \lambda p,$$

and

$$\lambda = \frac{q + r}{p}.$$

Similarly, setting

$$\frac{BP}{PE} = \mu, \quad \text{and} \quad \frac{CP}{PF} = \nu,$$

we obtain

$$\mu = \frac{p + q}{r} \quad \text{and} \quad \nu = \frac{r + p}{q}.$$

Now for any positive numbers x and y, the $A.M. - G.M.$ inequality gives

$$\frac{\frac{x}{y} + \frac{y}{x}}{2} \geq \sqrt{\frac{x}{y} \cdot \frac{y}{x}} = 1,$$

implying the well known result

$$\frac{x}{y} + \frac{y}{x} \geq 2;$$

similarly, we have the equally well known inequality

$$\frac{x+y}{2} \geq \sqrt{xy}, \quad \text{or} \quad x+y \geq 2\sqrt{xy}.$$

Hence

$$\frac{AP}{PD} + \frac{BP}{PE} + \frac{CP}{PF} = \lambda + \mu + \nu$$

$$= \frac{q+r}{p} + \frac{p+q}{r} + \frac{r+p}{q}$$

$$= \left(\frac{p}{q} + \frac{q}{p}\right) + \left(\frac{q}{r} + \frac{r}{q}\right) + \left(\frac{r}{p} + \frac{p}{r}\right)$$

$$\geq 2 + 2 + 2 = 6,$$

and

$$\frac{AP}{PD} \cdot \frac{BP}{PE} \cdot \frac{CP}{PF} = \lambda \cdot \mu \cdot \nu$$

$$= \frac{q+r}{p} \cdot \frac{p+q}{r} \cdot \frac{r+p}{q}$$

$$\geq \frac{2\sqrt{qr}}{p} \cdot \frac{2\sqrt{pq}}{r} \cdot \frac{2\sqrt{rp}}{q} = 8. \blacksquare$$

Three similar results are

(a) $\quad \dfrac{AD}{PD} \cdot \dfrac{BE}{PE} \cdot \dfrac{CF}{PF} \geq 27,$

(b) $\quad \dfrac{AD}{AP} + \dfrac{BE}{BP} + \dfrac{CF}{CP} \geq \dfrac{9}{2},$

and

(c) $\quad \dfrac{PD}{AP} + \dfrac{PE}{BP} + \dfrac{PF}{CP} \geq \dfrac{3}{2}.$

Finally, we observe that, for $k > 0$,

$$\frac{\lambda^k + \mu^k + \nu^k}{3} \geq (\lambda^k \mu^k \nu^k)^{1/3},$$

and since $\lambda \mu \nu \geq 8$, we have $\lambda^k + \mu^k + \nu^k \geq 3 \cdot 2^k$.

3. Next, let's consider four interesting little results due to John Rigby. We begin with a definition:

Let D, E, F be the feet of concurrent cevians from vertices A, B, C of $\triangle ABC$ (Figure 175(a)); we shall call $\triangle DEF$ a *cevian triangle*.

THEOREM 1. *If DEF is a cevian triangle of $\triangle ABC$, then the triangle $D'E'F'$, obtained by reflecting D, E, and F in the midpoints of their respective sides, is also a cevian triangle* (Figure 175(b)).

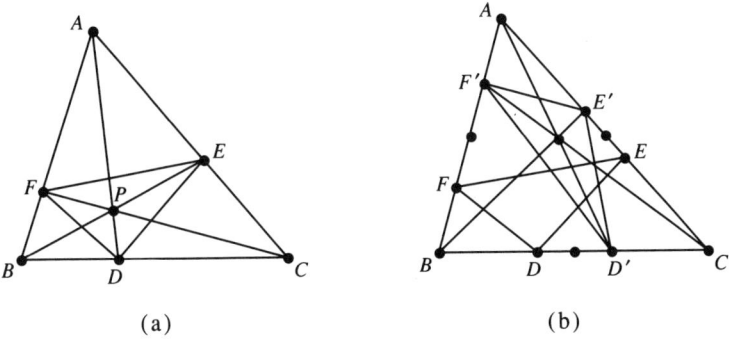

Figure 175

PROOF. By Ceva's theorem, DEF is a cevian triangle if and only if the product of the ratios in which $D, E,$ and F divide their respective sides is 1. Since reflection in the midpoint of a side merely *inverts* the ratio in which a point divides a side, the product of the ratios determined by $D', E',$ and F' is the reciprocal of 1, implying that $D'E'F'$ is indeed a cevian triangle. ∎

THEOREM 2. *If DEF is a cevian triangle of $\triangle ABC$, then the circumcircle of $\triangle DEF$ crosses the sides of $\triangle ABC$ at the vertices $D', E',$ and F' of another cevian triangle.*

PROOF. Let x, y, s and t denote the products

$$x = AF \cdot BD \cdot CE, \qquad y = FB \cdot DC \cdot EA,$$
$$s = AF' \cdot BD' \cdot CE', \qquad t = F'B \cdot D'C \cdot E'A.$$

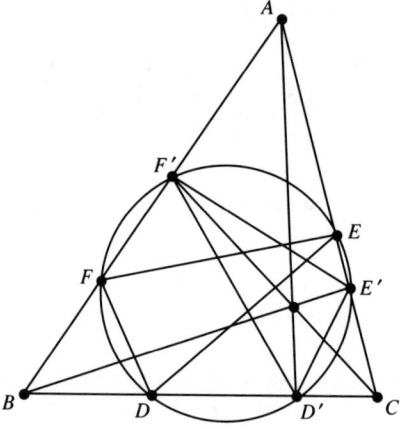

Figure 176

Since DEF is a cevian triangle, $x = y$; to prove that $D'E'F'$ is a cevian triangle we need to show that $s = t$.

The following argument needs a slight adjustment if any of D', E', F' occurs externally on a side. Recall that the product of a secant of a circle and the part outside the circle is the same for all secants from an external point. For the case illustrated in Figure 176, the secants from the vertices yield

$$AF' \cdot AF = EA \cdot E'A,$$
$$BD' \cdot BD = FB \cdot F'B,$$
$$CE' \cdot CE = DC \cdot D'C.$$

The product of these results gives $sx = ty$, and since $x = y$, it follows that $s = t$. ∎

Next we introduce a second definition:

if the pedal triangle of a point P in $\triangle ABC$ is a cevian triangle, the point P is called a pedal-cevian point, or PC-point for short.

Thus, for example, the circumcenter O of a triangle is always a PC-point since the vertices of its pedal triangle are the midpoints of the sides and the related cevians are just the medians which meet at the centroid G (Figure 177). Again, the vertices of the pedal triangle of the orthocenter H are the feet of the altitudes, which are also the related cevians, and since the altitudes meet at H itself, H is a special self-related PC-point. Recall also that the pedal triangle of the incenter I is the Gergonne triangle of $\triangle ABC$, and the cevians to its vertices concur at the Gergonne point (see Chapter 7). Now we easily establish two properties of PC-points.

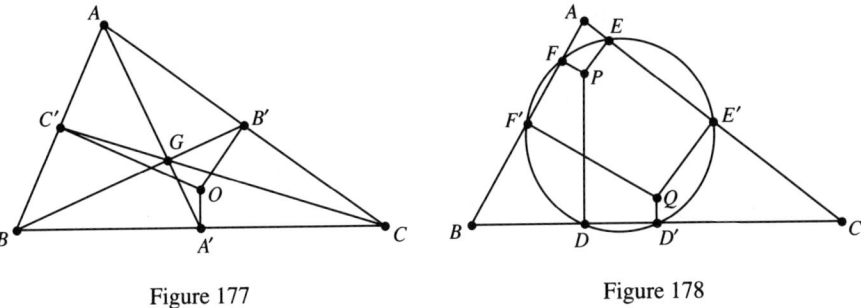

Figure 177   Figure 178

THEOREM 3. *If P is a PC-point of $\triangle ABC$, so is its isogonal conjugate Q.*

PROOF. This is an immediate consequence of Theorem 2 and the fact that the pedal triangles of a pair of isogonal conjugates (P, Q) have a common circumcircle, their *pedal circle* (Figure 178). (The pedal circle was introduced in Chapter 7, p. 67 (see also Figure 84, p. 68).

THEOREM 4. *If P is a PC-point of $\triangle ABC$, its reflection P' in the circumcenter O is also a PC-point.*

PROOF. Reflecting P in O and dropping perpendiculars to the sides from the image P' yields feet D', E', and F' which are respectively the reflections of D, E, and F in the midpoints of the sides (Figure 179). Thus, by Theorem 1, the pedal triangle $D'E'F'$ of P' is a cevian triangle and hence P' is a PC-point. ∎

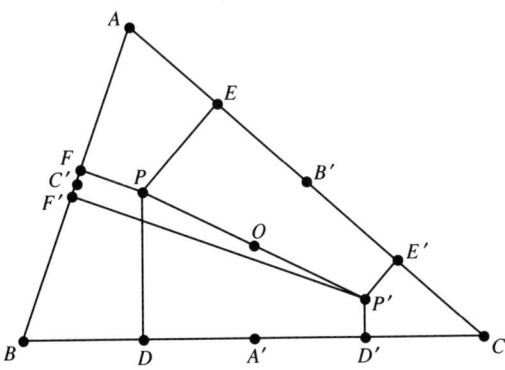

Figure 179

4. Haruki's Cevian theorem for circles

Finally, we conclude this essay with a delightful observation of Professor Hiroshi Haruki, a long-time colleague at the University of Waterloo until his retirement a few years ago.

HARUKI'S THEOREM. Suppose each of three circles intersects each of the others in two points; then, labelling the line segments as in Figure 180, we always have

$$\frac{a}{b} \cdot \frac{c}{d} \cdot \frac{e}{f} = 1.$$

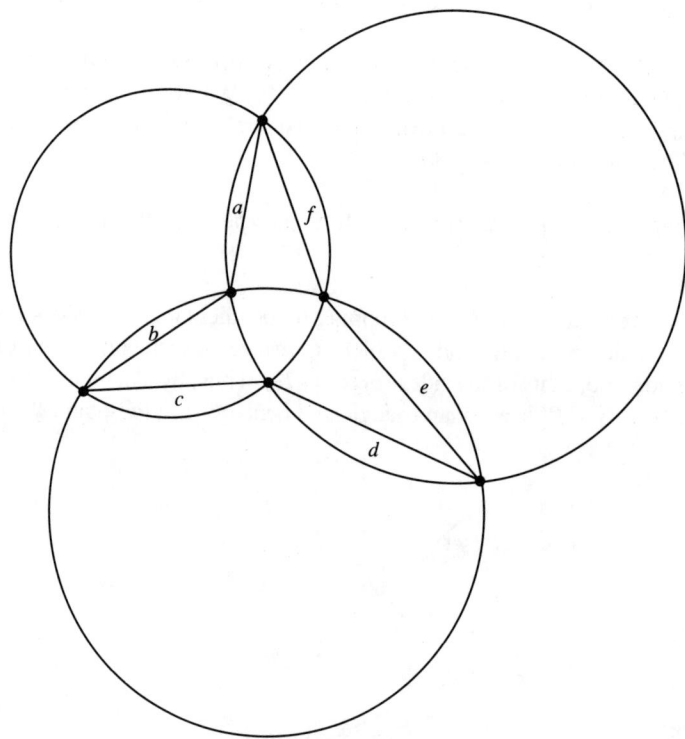

Figure 180

The proof is based on the fact that the common chord of two intersecting circles is the *radical axis* of the circles and that the three radical axes of a set of three circles, taken in pairs, are concurrent at a point called the *radical center*.

ON CEVIANS

[For full discussions of radical axes and centers, see e.g. *Geometry Revisited* by H.S.M. Coxeter and S. Greitzer, NML vol. 19, MAA 1967.]

In the region DEF common to the three circles, the three chords, each common to a pair of circles, cross at the radical center R to form six little segments (Figure 181). Let the three segments RD, RE, and RF have lengths x, y, and z.

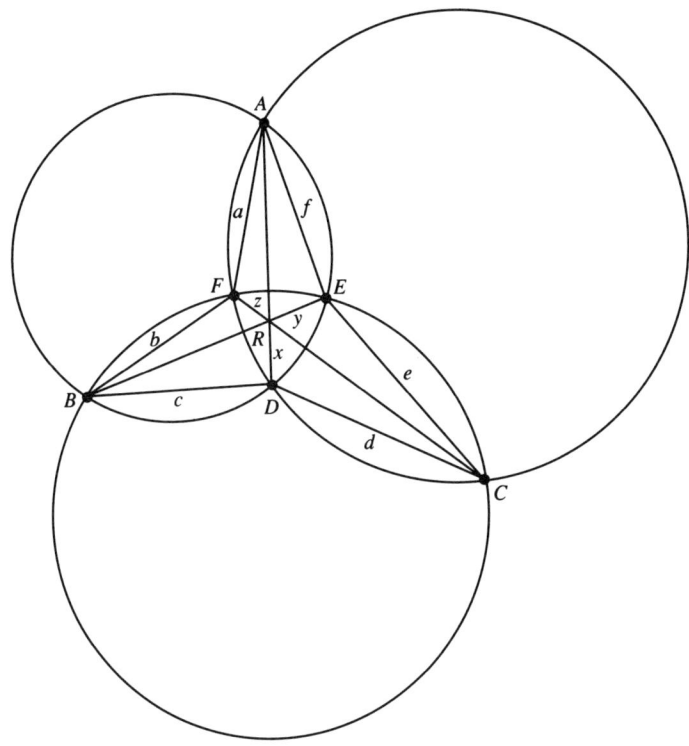

Figure 181

Now each circle contains two common chords intersecting at R to form a pair of similar triangles which contain among their sides two of a, b, c, d, e, f and two of x, y, z. For example, the case of the common chords BE and CF is illustrated in Figure 182; from the similar triangles, we obtain the proportion

$$\frac{b}{z} = \frac{e}{y}.$$

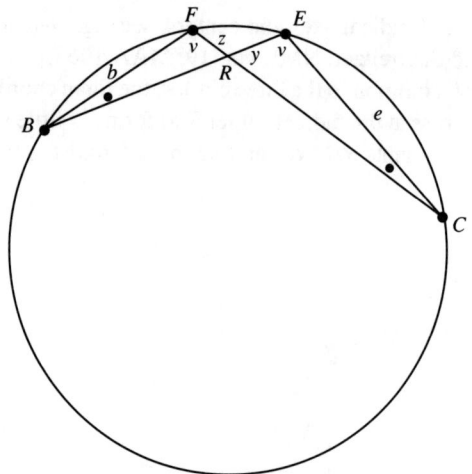

Figure 182

Altogether we get the three proportions

$$\frac{b}{z} = \frac{e}{y}, \quad \frac{f}{y} = \frac{c}{x}, \quad \frac{d}{x} = \frac{a}{z};$$

their product gives

$$\frac{bfd}{zyx} = \frac{eca}{yxz}, \quad \text{or} \quad bfd = eca,$$

from which the desired relation

$$\frac{a}{b} \cdot \frac{c}{d} \cdot \frac{e}{f} = 1$$

follows. ∎

CHAPTER THIRTEEN

The Theorem of Menelaus

1. In 1678, when Giovanni Ceva published his beautiful theorem, he also resurrected a companion theorem, due to Menelaus of Alexandria, that had virtually been forgotten by everybody since the first century A.D. While Ceva's theorem gives a necessary and sufficient condition for the *concurrence* of three *lines*, one through each *vertex* of a triangle, the theorem of Menelaus concerns the dual proposition, regarding the *collinearity* of three *points*, one on each *side* of a triangle. In both theorems the sole concern is the product of the ratios into which the sides are divided. The only difference is that, in the case of Menelaus, the product is −1 instead of +1.

THE THEOREM OF MENELAUS
The points L, M, and N on the sides BC, CA, and AB, respectively, of △ABC (Figure 183) are collinear if and only if the ratios of directed segments

$$\frac{AN}{NB} \cdot \frac{BL}{LC} \cdot \frac{CM}{MA} = -1$$

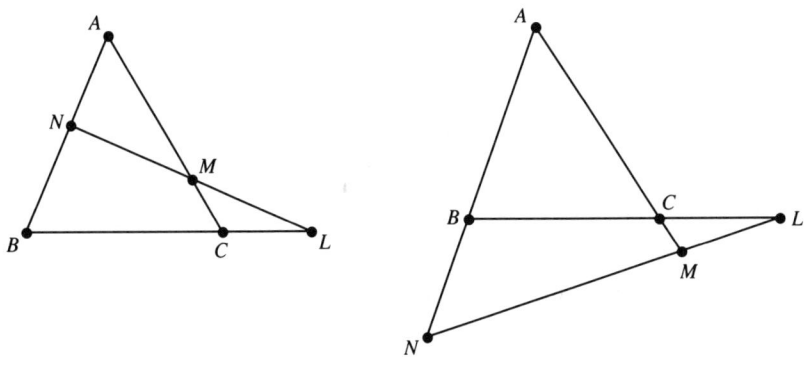

Figure 183

PROOF. Since a straight line cannot meet all three sides of a triangle internally, collinearity is out of the question unless at least one of the given points divides its side externally. Clearly, a straight line that meets two sides of a triangle externally must also meet the third side in the same manner. Hence, for collinearity, either one or all three of the divisions must be external, making the product of the corresponding ratios negative.

Necessity: Suppose L, M, and N *are* collinear (Figure 184).

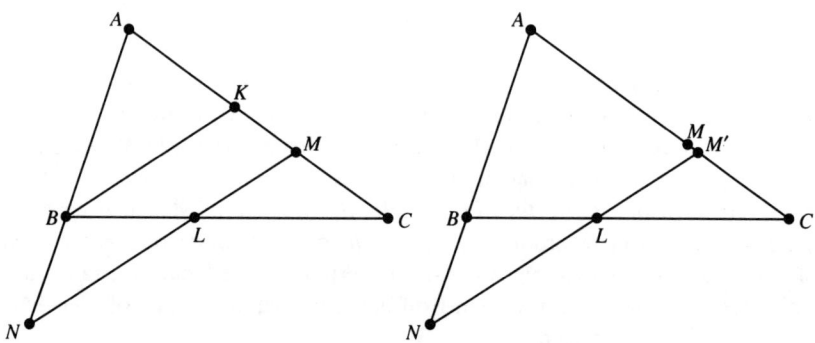

Figure 184   Figure 185

The key here is to draw BK parallel to NLM, getting the proportions

$$\frac{AN}{NB} = \frac{AM}{MK}, \quad \frac{BL}{LC} = \frac{KM}{MC}.$$

Then

$$\frac{AN}{NB} \cdot \frac{BL}{LC} \cdot \frac{CM}{MA} = \frac{AM}{MK} \cdot \frac{KM}{MC} \cdot \frac{CM}{MA} = \frac{AM}{MA} \cdot \frac{KM}{MK} \cdot \frac{CM}{MC} = (-1)^3 = -1.$$

Sufficiency: On the other hand, suppose

$$\frac{AN}{NB} \cdot \frac{BL}{LC} \cdot \frac{CM}{MA} = -1.$$

If NL crosses AC at M' (Figure 185), then, by the "necessity" just proved, we have

$$\frac{AN}{NB} \cdot \frac{BL}{LC} \cdot \frac{CM'}{M'A} = -1 = \frac{AN}{NB} \cdot \frac{BL}{LC} \cdot \frac{CM}{MA},$$

and
$$\frac{CM'}{M'A} = \frac{CM}{MA}.$$

Since there is only *one* point that divides a segment in a given ratio, M and M' must be the same point, and we conclude that L, M, and N are collinear. ∎

Now let's use the theorem to prove some unexpected properties of triangles.

2. Applications

(i) At each vertex of a triangle there is a bisector of the interior angle and also a bisector of the exterior angle, so that a triangle has six angle-bisectors altogether. Making allowances for the case of parallel lines, generally each bisector can be extended to determine a point on the *opposite* side. It strikes me as most remarkable that

> *the points L, M, and N, determined on the opposite sides of △ABC by an angle-bisector from each vertex, lie on a straight line provided*
> *either (a) all or (b) exactly one of the bisectors is external* (Figure 186(a) and (b)).

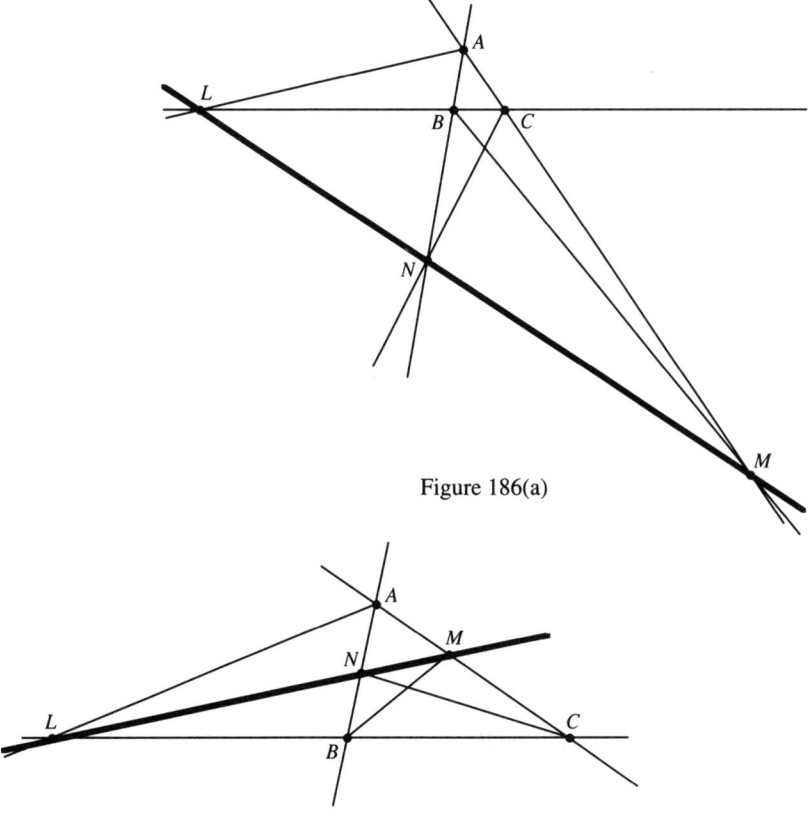

Figure 186(a)

Figure 186(b)

It is well known that the interior bisector divides the opposite side in the ratio of the sides about the bisected angle. The same proof yields the same conclusion for an external bisector (Figure 187).

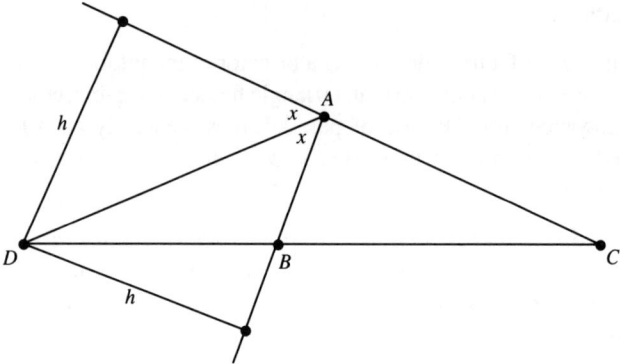

Figure 187

As is customary, denote the lengths of sides $BC, CA,$ and AB of $\triangle ABC$ by $a, b,$ and c. The point D on the bisector is equidistant from the arms of the angle, so we have

$$\frac{DB}{DC} = \frac{\text{Area}(\triangle ABD)}{\text{Area}(\triangle ADC)} = \frac{\frac{1}{2}AB \cdot h}{\frac{1}{2}AC \cdot h} = \frac{AB}{AC} = \frac{c}{b}.$$

Taking into account the orientation of the segments, we have

$$\frac{BD}{DC} = -\frac{DB}{DC} = -\frac{AB}{AC} = -\frac{c}{b}.$$

Therefore in case (a) above (all 3 bisectors external, Figure 186(a)),

$$\frac{AN}{NB} \cdot \frac{BL}{LC} \cdot \frac{CM}{MA} = \left(-\frac{b}{a}\right) \cdot \left(-\frac{c}{b}\right) \cdot \left(-\frac{a}{c}\right) = -1,$$

and in case (b) (only one bisector external, Figure 186(b))

$$\frac{AN}{NB} \cdot \frac{BL}{LC} \cdot \frac{CM}{MA} = \left(\frac{b}{a}\right) \cdot \left(-\frac{c}{b}\right) \cdot \left(\frac{a}{c}\right) = -1,$$

and the conclusions are established immediately by the theorem of Menelaus.

When all six angle bisectors are considered, all possibilities of (a) either those from all vertices being external, or (b) exactly one being external, consist of these four cases:

I A_e, B_e, C_e;   II A_i, B_e, C_i;   III A_i, B_i, C_e;   IV A_e, B_i, C_i.

[Here A_i, A_e, for example, denote the internal and external angle bisector, respectively, from A to the side opposite.] Figure 188 illustrates these four sets of collinear points lying on lines I, II, III and IV, respectively.

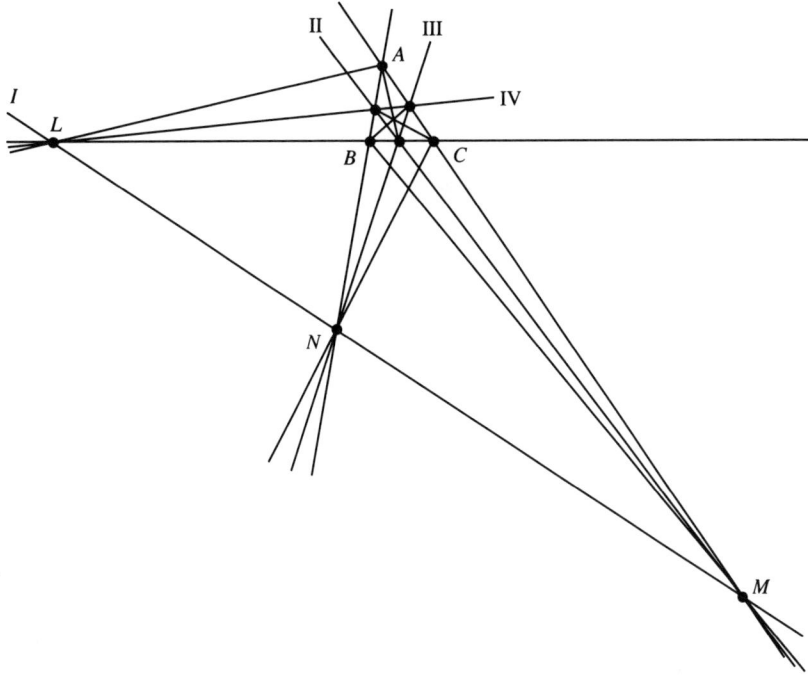

Figure 188

(ii) THE ORTHIC AXIS

Let the orthic triangle of $\triangle ABC$ be $\triangle DEF$, where D, E, F are the feet of the altitudes from A, B, C, respectively (Figures 189(a), (b)). Now each side of each triangle meets the three sides of the other triangle, possibly externally; let the non-vertex points of intersection be L, M, N. Then

the points L, M, N lie on a straight line called the orthic axis of $\triangle ABC$.

Easily established cyclic quadrilaterals and vertical angles yield five equal angles y, as shown in Figure 189(c), (which illustrates the case of an acute angled triangle), revealing that BC is the exterior angle bisector at D of the orthic triangle, and leading to the general result that *the sides of a triangle are the exterior angle bisectors of its orthic triangle.* The orthic axis, then, is simply a particular case of part (a) (three external bisectors) of the previous result. ∎

(iii) Again allowing for the case of parallel lines,

the tangents to the circumcircle of a triangle at the vertices meet the opposite sides in three collinear points (Figure 190).

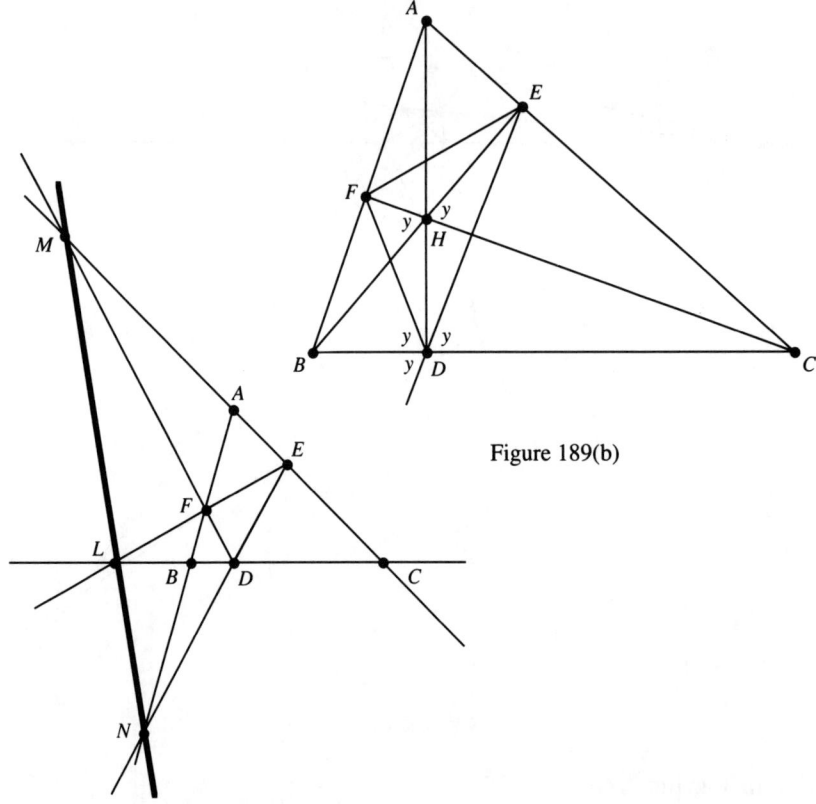

Figure 189(b)

Figure 189(a)

Consider the tangent AL and the secant LBC. Because the angle between a tangent and a chord is equal to the peripheral angle subtending that chord, the triangles LBA and LAC are similar. From the proportional sides, then, we get

$$\frac{LB}{LA} = \frac{LA}{LC} = \frac{AB}{AC} = \frac{c}{b}.$$

Therefore

$$\frac{LB}{LC} = \frac{LB}{LA} \cdot \frac{LA}{LC} = \frac{c^2}{b^2},$$

making

$$\frac{BL}{LC} = -\frac{c^2}{b^2}.$$

THE THEOREM OF MENELAUS

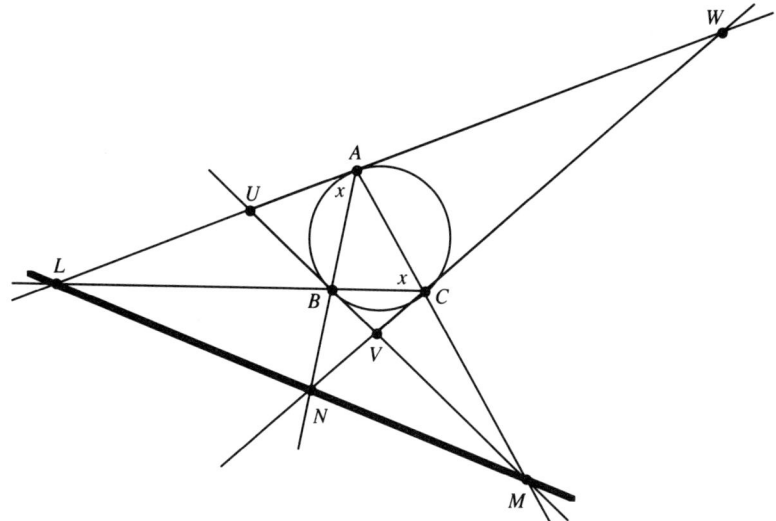

Figure 190

Similarly,

$$\frac{CM}{MA} = -\frac{a^2}{c^2}, \quad \text{and} \quad \frac{AN}{NB} = -\frac{b^2}{a^2},$$

giving

$$\frac{AN}{NB} \cdot \frac{BL}{LC} \cdot \frac{CM}{MA} = -1,$$

from which the desired collinearity follows by the theorem of Menelaus. ∎

(iv) Finally, *Suppose L, M, and N are points on the sides of △ABC or their extensions, one point on each side, and that each of the points is reflected in the midpoint of its side to give the respective images L', M', and N'* (Figure 191). *Then L', M', and N' are collinear if and only if L, M, and N are collinear.*

The proof is immediate for, as we have observed before, the images L', M', and N' divide the sides in ratios that are the reciprocals of the ratios determined by L, M, and N. By the theorem of Menelaus, L', M', and N' are collinear if and only if the product of their associated ratios is -1. Thus, L', M', and N' are collinear

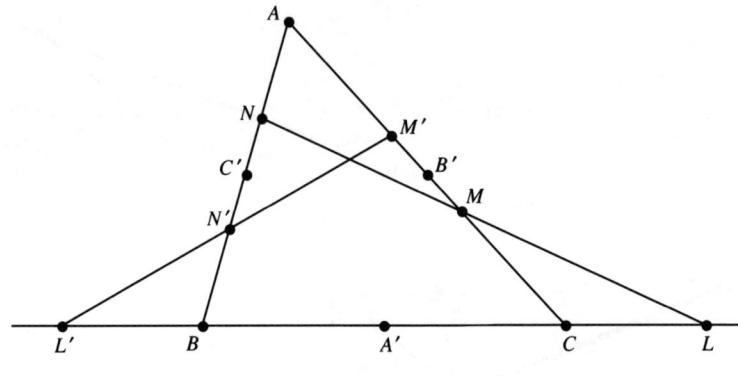

Figure 191

if and only if the corresponding product for L, M, and N is the reciprocal of -1, which is again just -1, and which is the case if and only if L, M, and N are themselves collinear. ∎

Suggested Reading

Instead of attempting a long list of the many excellent standard works on geometry, I would like to mention, in addition to the three outstanding books ([1], [2], [3] below) only four unusual books of self-contained topics that can be enjoyed in one sitting:

[1] Nathan Altshiller Court, *College Geometry* (Barnes and Noble, 1925).
[2] Roger Johnson, *Advanced Euclidean Geometry* (Dover, 1929).
[3] H. S. M. Coxeter and S. Greitzer, *Geometry Revisited*, New Mathematical Library, vol. 19 (MAA, 1967).

1. E. A. Maxwell, *Geometry For Advanced Pupils* (Oxford, 1953).
2. E. H. Lockwood, *A Book of Curves* (Cambridge, 1961).
3. H. Fukagawa and D. Pedoe, *Japanese Temple Geometry Problems* (The Charles Babbage Research Centre, 1989).
4. H. M. Cundy and A. P. Rollet, *Mathematical Models* (Oxford, 1960).

You might also enjoy the following article, which can be read in three minutes: Dwight Paine, Triangle Rhyme, *Mathematics Magazine*, 56 (1983), 235–238.

Solutions to the Exercises

1. Cleavers and Splitters

1. Since Q_1 and Q_2 bisect AM and BM, Q_1Q_2 is parallel to AB, as is $A'B'$ (Figure 15). Hence $Q_1Q_2 \parallel A'B'$, and similarly the three sides of $\triangle Q_1Q_2Q_3$ are respectively parallel to those of $\triangle A'B'C'$.

   Since these triangles have a common incircle with center S, the half-turn about S would carry the one into the other, confirming the conjecture: Q_1Q_2 would run along $A'B'$, and Q_1Q_3 along $A'C'$, forcing Q_1 to go to A', etc. ∎

2. Let the incircles of triangles ABX and ACX touch AX respectively at P and Q (Figure 16). We shall show that $P = Q$ by showing that $XP = XQ$.

   Let s, s_1, and s_2 denote the semiperimeters of triangles ABC, ABX, and ACX. Then, referring to Figure 16, we have the standard relations
   $$BX = s - b, \quad \text{and} \quad CX = s - c;$$
   in triangles ABX and AXC, the corresponding relations are
   $$XP = s_1 - AB = s_1 - c, \quad \text{and} \quad XQ = s_2 - AC = s_2 - b$$
   (Q is not marked, of course, in this figure). Hence
   $$XP - XQ = s_1 - s_2 + b - c.$$

   Now,
   $$2s_1 = c + BX + AX \quad \text{and} \quad 2s_2 = AX + XC + b.$$

   Thus
   $$2(s_1 - s_2) = c - b + BX - XC = c - b + (s - b) - (s - c)$$
   $$= 2c - 2b,$$
   so that
   $$s_1 - s_2 = c - b.$$

   Hence
   $$XP - XQ = c - b + b - c = 0,$$
   as claimed. ∎

   The case of excircles can be handled similarly.

3. (i) The Logic of the Exercise:

From (a), masses of a, b, c at A, B, C are equivalent to a mass of $a + b + c$ at I; from (b), masses of $2s - 2a$, $2s - 2b$, $2s - 2c$ at A, B, C are equivalent to a mass of

$$(2s - 2a) + (2s - 2b) + (2s - 2c) = 6s - 4s = 2s = a + b + c \text{ at } M.$$

Since these systems have the same total mass, a system consisting of *both* these systems would have its center of mass at the midpoint of IM. But such a compound system would have masses of

$$a + (2s - 2a) = b + c \quad \text{at } A,$$
$$b + (2s - 2b) = c + a \quad \text{at } B,$$
and
$$c + (2s - 2c) = a + b \quad \text{at } C.$$

Thus part (c) identifies this midpoint of IM as S.

(ii) It remains to establish claims (a), (b), and (c). However, part (a) follows easily with an argument analogous to that associated with Figure 5 in Section 1(d) on the Spieker circle, and (b) similarly follows from the discussion on the Nagel point in Section 2(a) (Figures 6 and 7). We shall leave these routine matters and close the solution with a proof of (c).

We have seen in Section 1(d) that masses of a, b, c at the respective midpoints A', B', C' of the sides have center of gravity at S. Doubling all the masses in a system does not change the center of gravity, and therefore masses of $2a, 2b, 2c$ at A', B', C' also have center of gravity at S. Now, since A' is the midpoint of BC, a mass of $2a$ at A' is equivalent to a mass of a at each of B and C; similarly for the masses at B' and C', and it follows that the center of gravity of masses $b + c$ at A, $c + a$ at B, and $a + b$ at C also have center of gravity at S (see the figure below). ∎

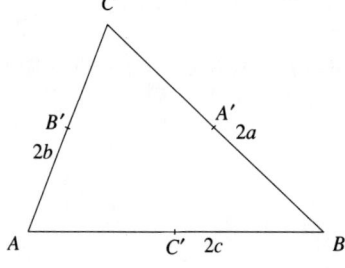 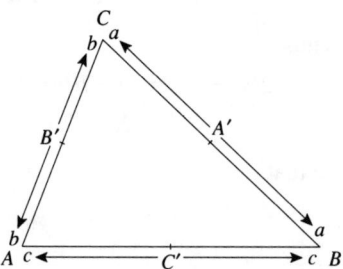

4. Model a solution after the solution of Exercise 3.

SOLUTIONS TO THE EXERCISES   159

2. The Orthocenter

Work backwards by extending BO and CO to meet the circle at P and Q, thus forming rectangle $QBCP$ (Figure 192). Let QP cross AB at U and AC at V, and let perpendiculars from U and V meet BC at L and M. We want to prove that $LH \parallel AB$ and $HM \parallel AC$. Since these cases have similar proofs, we show only that $LH \parallel AB$.

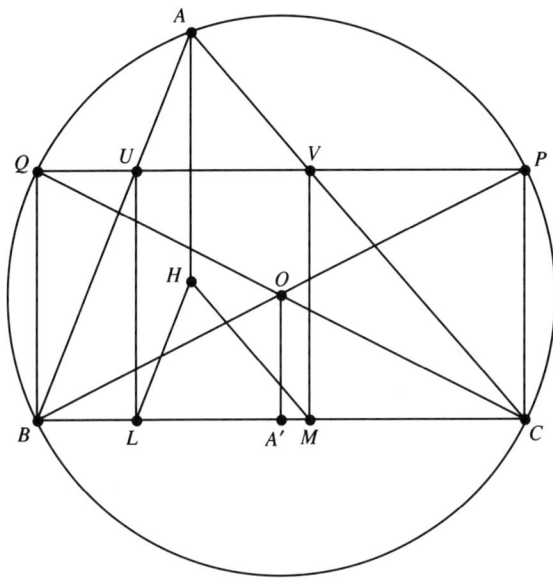

Figure 192

Since O and A' are midpoints of sides QC and BC in $\triangle QBC$, we have $QB = 2 \cdot OA'$, which in turn is equal to AH (established in the text), making QB and AH equal and parallel and therefore $QBHA$ a parallelogram. But $QBLU$ is a rectangle. Thus AH and UL are equal and parallel, making $ULHA$ a parallelogram and $LH \parallel AU$, that is, $LH \parallel AB$. ∎

3. On Triangles

1. Let K be the intersection of the perpendicular bisector of AU and the circumdiameter AOD. We shall show that KU is perpendicular to BC.

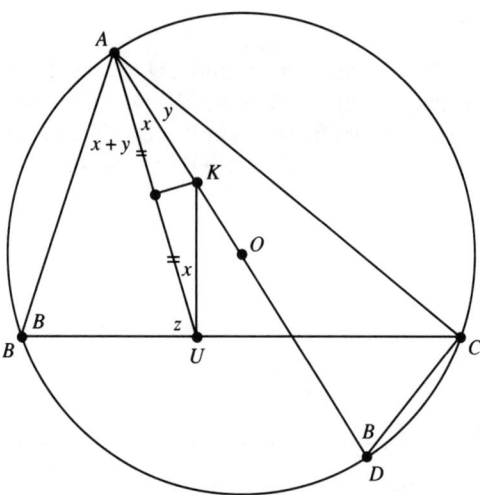

Figure 193

In Figure 193, $KA = KU$, giving $\angle KAU = \angle KUA$ ($= x$) in $\triangle KAU$. Let $\angle KAC = y$; then since AU bisects angle A, we have $\angle BAU = x + y$. Also, let $\angle AUB = z$. We shall show that $x + z = 90°$ at U.

Clearly $\angle B = \angle D$ on chord AC, and since AD is a diameter, then $\angle ACD = 90°$. Hence, in $\triangle ADC$, $B + y = 90°$. Thus, in $\triangle ABU$, we have

$$B + y + x + z = 180°,$$

giving the desired

$$x + z = 90°. \blacksquare$$

2. Since X and A' are the midpoints of BE and BC, we have $XA' \parallel EC$, making XA' perpendicular to BE (Figure 40). Thus the circle on HA' as diameter goes through X. Similarly, this circle goes through Y, and since $\angle HDA'$ is a right angle, it also goes through D. Thus the first part of the problem is settled.

Right triangles ADC and AEH have a common angle at A, implying $\angle C = \angle AHE$. But $\angle AHE = \angle XHD$ (vertically opposite) $= \angle XYD$ (in the same segment of the circle), and we have $\angle C$ in $\triangle ABC$ equal to $\angle Y$ in $\triangle XYD$. Similarly $\angle B = \angle DXY$, implying the triangles are similar. $\blacksquare$

4. On Quadrilaterals

1. Referring to Figure 48, since T is the anticenter, both ST and QT are altitudes of $\triangle ESQ$, making T the orthocenter of the triangle. Thus the third altitude, from E, also goes through T.

SOLUTIONS TO THE EXERCISES   161

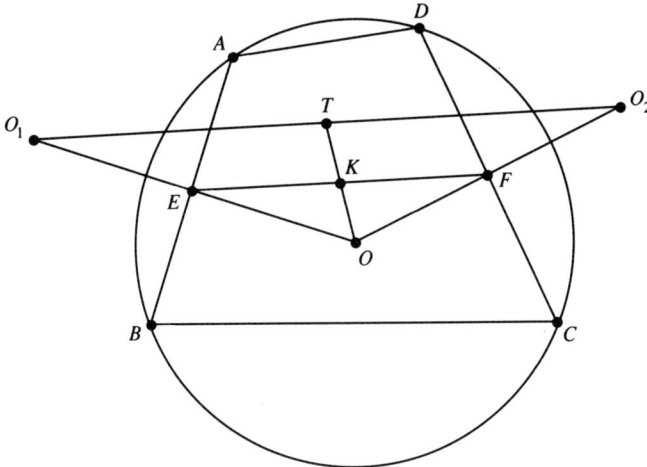

Figure 194

2. Clearly OO_1 and OO_2 cross AB and CD at their midpoints E and F (Figure 194). Hence the centroid K lies on EF (actually the midpoint). But E and F are also the midpoints of OO_1 and OO_2. Therefore the dilatation $O(2)$ carries E to O_1, F to O_2, and K to some point on O_1O_2. But, by the definition of the anticenter T, $O(2)$ takes K into T. Hence T lies on O_1O_2. ∎

3. Let the sides of $ABCD$ be a, b, c, d (Figure 195(a)). Suppose the bisectors of angles B and D meet at P on AC. Then

$$\frac{a}{b} = \frac{AP}{PC} = \frac{d}{c},$$

giving

$$\frac{a}{d} = \frac{b}{c}.$$

Let the bisector of angle A meet BD at Q (Figure 195(b)). We shall show that CQ bisects angle C. Because AQ bisects angle A,

$$\frac{a}{d} = \frac{BQ}{QD}.$$

But, since

$$\frac{a}{d} = \frac{b}{c}, \quad \text{we have} \quad \frac{BQ}{QD} = \frac{b}{c},$$

implying CQ bisects angle C. ∎

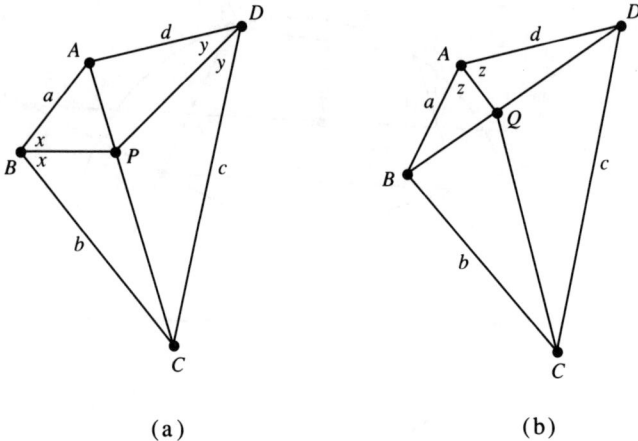

Figure 195

7. The Symmedian Point

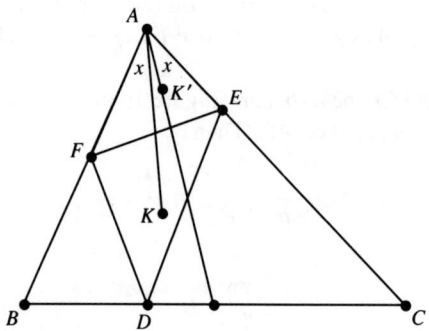

Figure 196

1. Recall that the symmedians bisect the sides of the orthic triangle. Hence the symmedian AK of $\triangle ABC$ lies along the median of $\triangle AFE$ from A. With regard to $\triangle AFE$, then, AK and AK' are median and symmedian at A, and accordingly, are a pair of isogonal conjugates (Figure 196). Hence in $\triangle ABC$, AK' is isogonal to symmedian AK, making it the median from A. ∎

SOLUTIONS TO THE EXERCISES 163

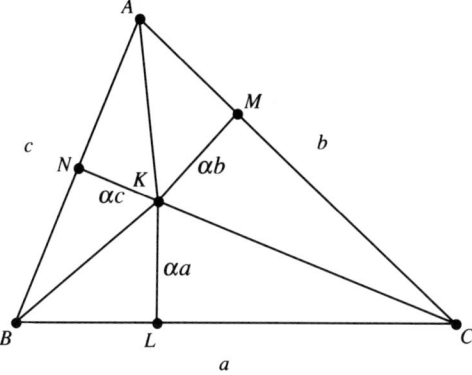

Figure 197

2. Recall that the perpendicular distances from K to the sides of the triangle are proportional to the lengths of the sides. Hence for some real number α, $KL = \alpha a$, $KM = \alpha b$, and $KN = \alpha c$ (Figure 197). Now lines from K to the vertices partition $\triangle ABC$ into three triangles the sum of whose areas yields

$$\frac{1}{2} a \cdot \alpha a + \frac{1}{2} b \cdot \alpha b + \frac{1}{2} c \cdot \alpha c = \Delta,$$

which immediately gives the desired value α. ∎

3. In the identity

$$(x^2 + y^2 + z^2)(a^2 + b^2 + c^2) - (ax + by + cz)^2$$
$$= (bx - ay)^2 + (cy - bz)^2 + (cx - az)^2,$$

$a^2 + b^2 + c^2$ is a constant as P varies, and so is $ax + by + cz \, (= 2\Delta)$. Hence $S = x^2 + y^2 + z^2$ is a minimum when the right side is a minimum.

For P at the symmedian point K, we have

$$\frac{x}{y} = \frac{a}{b},$$

giving $bx - ay = 0$; similarly, $cy - bz = 0$ and $cx - az = 0$. $P = K$, then, makes the right side zero, which is clearly the minimum value of a sum of three squares. ∎

4. Let masses of a^2, b^2, and c^2 be suspended at the vertices A, B, and C, respectively (Figure 198). The center of mass of the masses at B and C will be a point X on BC about which their moments are equal:

$$b^2 \cdot BX = c^2 \cdot XC,$$

giving

$$\frac{BX}{XC} = \frac{c^2}{b^2}.$$

That is to say, X divides BC in the same ratio as the symmedian AP does. Hence $X = P$, and we conclude that the center of mass M of the whole system lies on the symmedian from A. Similarly, M lies on the other symmedians, and we have $M = K$, the symmedian point of $\triangle ABC$.

Hence the system is equivalent to masses of $(b^2 + c^2)$ at P and a^2 at A. Since the moments about K must be equal, we have

$$a^2 \cdot AK = (b^2 + c^2) \cdot KP,$$

giving

$$\frac{AK}{KP} = \frac{b^2 + c^2}{a^2}. \blacksquare$$

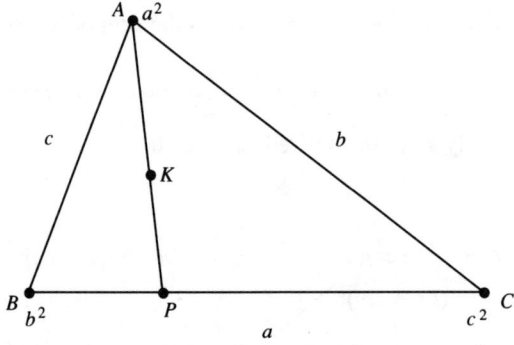

Figure 198

5. Because the centroid G divides a median in the ratio of 2:1, in Figure 199(b), GB' is half as long as BG, and therefore, extending GB' its own length to H makes G the midpoint of BH. Since C' is the midpoint of AB, then, in $\triangle ABH$, $C'G$ joins the midpoints of two of the sides and is parallel to AH and one-half as long. But $C'G$ is also one-half GC, and so AH is equal and parallel to GC. Thus the sides of $\triangle AGH$ are two-thirds as long as the medians of $\triangle ABC$, and its angles are the angles between the medians of $\triangle ABC$.

Turning to Figure 199(a), since AK and AG are isogonal conjugates at A, the line NM, joining the feet of the perpendiculars from K to the sides, is perpendicular to AG. Similarly, the other sides of $\triangle LMN$ are perpendicular to the other medians, implying that the angles in $\triangle LMN$ are the angles between

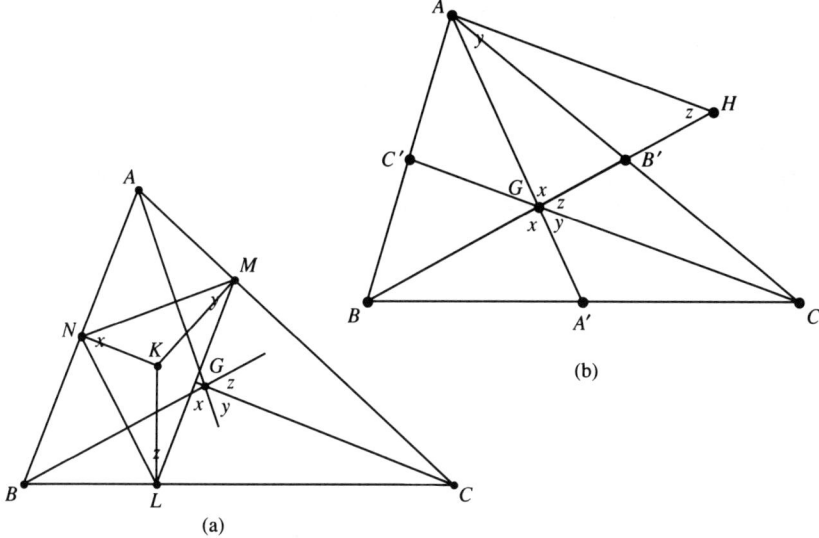

Figure 199

the medians. Hence $\triangle LMN$ in Figure (a) is similar to $\triangle HAG$ in Figure (b). Since the sides of $\triangle HAG$ are proportional to the medians of $\triangle ABC$, so are the sides of $\triangle LMN$. ∎

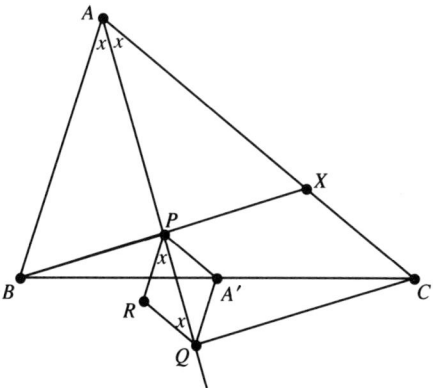

Figure 200

6. If BP is extended to meet AC at X, then $\triangle ABP \cong \triangle APX$, showing that P is the midpoint of BX (Figure 200). In $\triangle BCX$, then, PA' joins the midpoints of

two of the sides and is therefore parallel to *CX*. Thus *PA'* is also parallel to *RQ*. Similarly, *QA'* is parallel to *RP* and *PRQA'* is a parallelogram.

Also, corresponding angles x at *P* and *A*, and alternate angles x at *Q* and *A* make $\triangle PQR$ isosceles with $RP = RQ$, implying that *PRQA'* is in fact a rhombus. Since the diagonals of a rhombus bisect each other at right angles, the reflection of *AA'* in the angle bisector *APQ* would give *AR*. Since reflection in an angle bisector carries a median into a symmedian, the conclusion follows. ∎

PRELIMINARY WORK FOR #7 AND #8 (JOHN RIGBY).

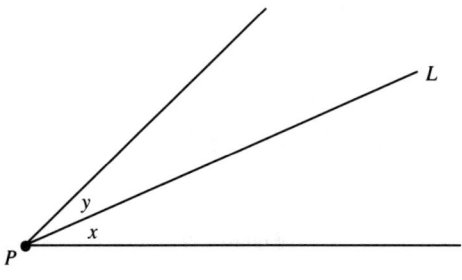

Figure 201

If a line *L* divides an angle *P* into parts x and y, let us describe the partition with the ordered pair (x, y) in which the parts are listed in the order they are encountered as the angle is traversed in the positive sense (Figure 201).

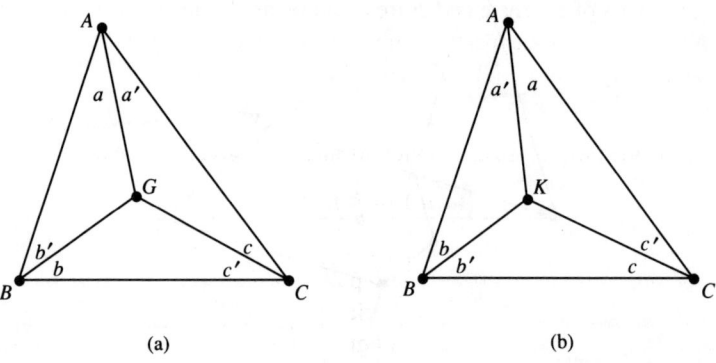

Figure 202

Suppose the medians of $\triangle ABC$ divide the angles through which they pass as described by the triple $[(a, a'), (b, b'), (c, c')]$ (Figure 202(a)). Since the symmedians are the isogonal conjugates of the medians, they will divide the angles as described by the reversed pairs $[(a',a),(b',b),(c',c)]$ (Figure 202(b)).

As in the solution to #5, let GB' be extended its own length to H to yield $\triangle AGH$, which has sides that are two-thirds as long as the medians of $\triangle ABC$ (Figure 203). Now let us proceed to show that the medians of $\triangle AGH$ are parallel to the sides of $\triangle ABC$.

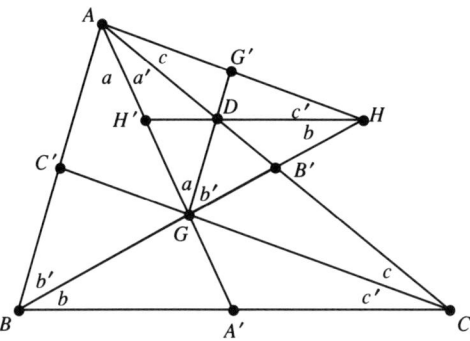

Figure 203

Clearly median AB' is parallel to side AC along which it lies. Median GG' joins the midpoints of sides BH and AH of $\triangle BHA$, so is parallel to the third side, AB. To see that median HH' is parallel to BC, observe that H' and D trisect AA' and AC. ∎

From pairs of alternate and corresponding angles in connection with these parallel lines, and other pertinent relations that are evident, it is easy to deduce the sizes of the other angles as marked in Figure 203.

Thus we have the following two results:

(i) If the medians of $\triangle ABC$ divide its angles as given by the triple

$$[(a,a'),(b,b'),(c,c')],$$

then, in a triangle T having angles of sizes $a' + c$, $b' + a$, and $c' + b$, the lines which divide them respectively into parts given by $[(a',c),(b',a),(c',b)]$ are in fact the medians of T. (All such triangles would be similar to $\triangle AHG$, and the partitioning lines would give a figure similar to $\triangle AGH$ with D inside it.)

Reversing the pairs throughout translates this result into the following property of symmedians rather than medians.

(ii) If the symmedians of a triangle ABC divide the angles as described by $[(a',a),(b',b),(c',c)]$, then, in a triangle T having angles equal to $a' + c$,

$b' + a$, and $c' + b$, the lines which respectively divide them into parts described by $[(c, a'), (a, b'), (b, c')]$ are the symmedians of T:

if the symmedians are described by $[(a', a), (b', b), (c', c)]$, then the medians of $\triangle ABC$ are given by $[(a, a'), (b, b'), (c, c')]$, and by result (i), we have that the medians of T are given by $[(a', c), (b', a), (c', b)]$, which in turn implies that the symmedians T are given by $[(c, a'), (a, b'), (b, c')]$. ∎

Now let's apply these results to Exercises 7 and 8.

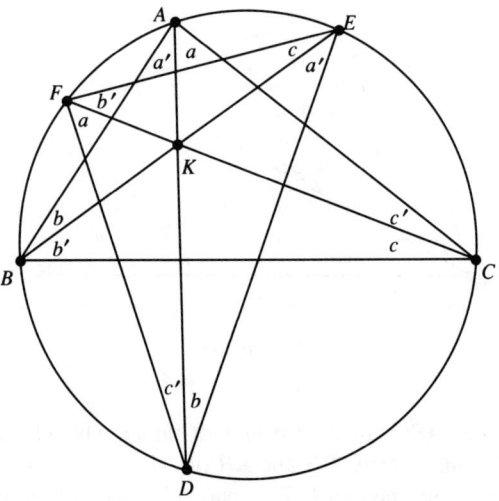

Figure 204

7. Let KA, KB, and KC divide the angles of $\triangle ABC$ as given by $[(a', a), (b', b), (c', c)]$ (Figure 204). Since angles in the same segment of a circle are equal, the angles at D, E, and F are partitioned as marked in the figure. Thus $\triangle EFD$ has angles $a' + c$, $b' + a$, and $c' + b$, which are partitioned by KE, KF, and KD as given by $[(c, a'), (a, b'), (b, c')]$. By preliminary result (ii), then, KE, KF, and KD are the symmedians of $\triangle DEF$ and K is its symmedian point. ∎

8. First let us show that G and O are a pair of isogonal conjugates in the triangle of centers PQR.

Since the line joining the centers of two intersecting circles is perpendicular to their common chord, PR and PQ are perpendicular to BG and CG, respectively. Also, OP is perpendicular to the common chord BC of circle I and the circumcircle of $\triangle ABC$ (Figure 205). In right triangles PTS and SCU, the equal vertically opposite angles at S give equal third angles $\angle TPS = \angle GCB$.

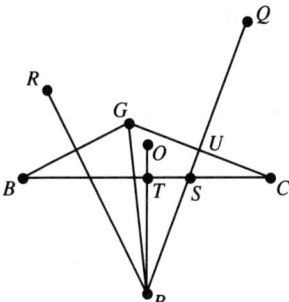

Figure 205

Now, referring to Figures 205 and 99 (page 77), circle I, with center P, goes through B, G, and C, and therefore $\angle BPG$ at the center is twice the angle BCG at the circumference. But PR bisects this angle BPG (the perpendicular PR to the chord BG from the center bisects the central angle BPG), and we have $\angle RPG = \angle GCB = \angle TPS$, which is $\angle QPO$, making GP and OP a pair of isogonal lines at P in $\triangle PQR$. Similarly, GQ and OQ are isogonal at Q, and GR and OR are isogonal at R, making G and O a pair of isogonal conjugates of $\triangle PQR$, as claimed. ∎

It follows, then, that if one of these points G and O is the centroid of $\triangle PQR$, the other must be its symmedian point. Let us conclude by showing that O is indeed the centroid of $\triangle PQR$.

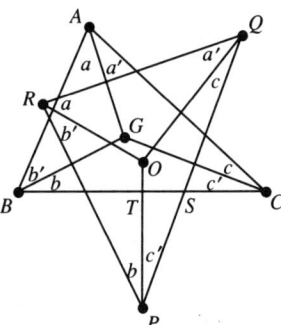

Figure 206

Suppose the medians of $\triangle ABC$ partition its angles as described by $[(a, a'), (b, b'), (c, c')]$ (Figure 206). Now, just as we proved that $\angle TPS = \angle GCB$ (= c' in the figure), so is $\angle TPR = \angle GBC$ (= b), and similarly around the figure, the angles occur at the other vertices Q and R as shown. Thus, in $\triangle PQR$, the lines OQ, OR, and OP partition its angles as given by $[(a', c), (b', a), (c', b)]$, and O is indeed its centroid by preliminary result (i). ∎

CONCLUDING REMARKS. Let X and Y be generic names for the points of interest in triangle ABC, like G and O, and let PQR be the triangle of centers of circles that pass through two of the vertices of $\triangle ABC$ and the point X. Then, we have just proved that

(a) If X and Y are the centroid G and the circumcenter O of $\triangle ABC$, then, in $\triangle PQR$, X is the symmedian point and Y is the centroid.

This ingenious approach to Exercises 7 and 8 is due to Dr. Rigby, who also pointed out the following additional results.

(b) If X and Y are the orthocenter H and circumcenter O of $\triangle ABC$, then, in $\triangle PQR$, X is the circumcenter and Y the orthocenter.

(c) If X and Y are the incenter I and circumcenter O of $\triangle ABC$, then, in $\triangle PQR$, X is the orthocenter and Y the circumcenter.

9. The Tucker Circles

1. An antiparallel PQ to BC makes $\angle APQ = \angle ACB = \gamma$ (Figure 207(a)). Since the angle between a tangent and a chord is equal to the angle in the segment on the opposite side of the chord, $\angle TAB$ is also equal to γ. Hence $\angle TAP = \angle APQ$, and PQ is parallel to AT.

2. We know, by construction, that $PS \parallel DE$, and we saw in Chapter 7 on the symmedian point that $PD = DQ = SE = ER$ (Figure 207(b)). Also, CD and CE are equal tangents to the incircle of $\triangle ABC$. Hence

$$CQ = CD - DQ = CE - ER = CR,$$

making $\triangle CRQ$ isosceles and $\angle QRC = \frac{1}{2}(180° - C)$. But, in isosceles triangle CDE, we also have $\angle DEC = \frac{1}{2}(180° - C)$. Hence $\angle DEC = \angle QRC$, and we have $QR \parallel DE$.

In this case, DE is the midline of the strip between the parallels PGS and QR, and as such it bisects every segment from G to QR. Thus the dilatation $G(2)$ carries DE to lie along the line of QR, i.e. along XY. Similarly, it carries FE to lie along ZY and FD to lie along ZX. In short, it carries $\triangle DEF$ into $\triangle XYZ$. Recall also that G is the symmedian point of the Gergonne triangle $\triangle DEF$. It follows that G is the symmedian point of the similar triangle XYZ (recall that, since G is the symmedian point, the distances from G to the sides of $\triangle DEF$ are proportional to the sides of $\triangle DEF$ themselves, and this property is preserved by any dilatation having center G). ∎

SOLUTIONS TO THE EXERCISES   171

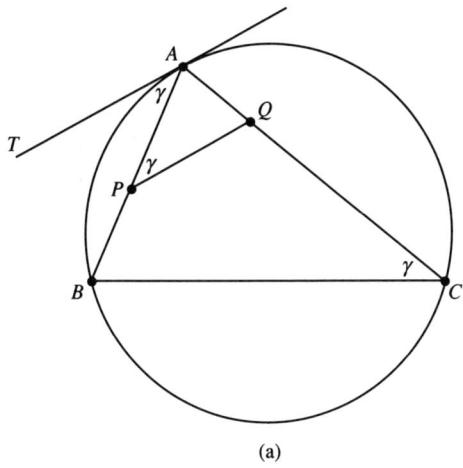

(a)

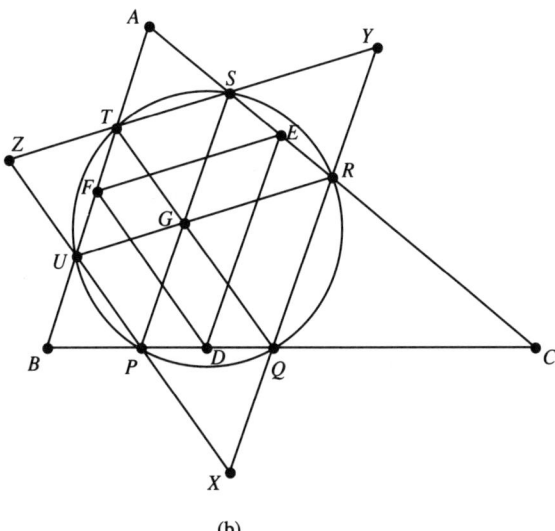

(b)

Figure 207

Since *PS*, *QT*, and *UR* are parallel to the sides of $\triangle DEF$, they are also respectively parallel to the sides of $\triangle XYZ$. Since they pass through the symmedian point *G*, the circle *PQRSTU* which they determine on the sides is its first Lemoine circle, as well as being the Adams circle of $\triangle ABC$. ∎

11. The Orthopole

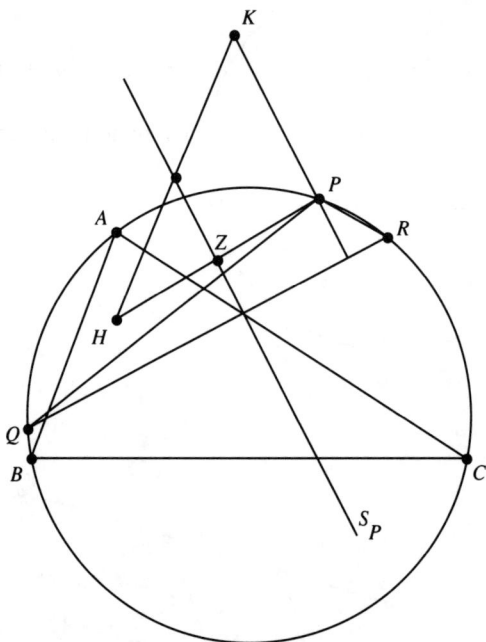

Figure 208

In Figure 208, the Simson line of P and the altitude of $\triangle PQR$ from P are both perpendicular to QR, and hence they are parallel. But in Chapter 5 on *A Property of Triangles*, we saw that the Simson line of any point P bisects the segment HP to the orthocenter of the reference triangle ABC. Thus, in $\triangle PKH$, the Simson line of P passes through the midpoint Z of HP, is parallel to PK, and therefore bisects HK. Similarly for the Simson lines of Q and R. Since the Rigby point X is the point of concurrence of these Simson lines, X must indeed be the midpoint of HK. ∎

Index

Adam's Circle, 62, 98
Angle Bisectors, 149
Anticenter, 36
Antiparallels, 85
Archimedes' Theorem, 1
Arithmetic Mean-
  Geometric Mean Inequality, 140

Bottema, O., 139
Boyer, Carl, 2
Brahmagupta's Theorem, 37
Brocard
  Angle, 101, 102
  Circle, 106
  Diameter, 121
  Points, 99
  Ray, 122
  Triangles, 110, 118, 121

Center of Gravity, 4
Centroid
  Quadrilateral, 36, 40
  Triangle, 7, 112
Ceva's Theorem, 61, 137
Cevian Triangle, 141
Cleavance-center, 2
Collinearity, 30
Concurrence, 31, 36, 116

Dilatations, 11, 12, 15, 30, 31, 64, 66, 67, 117
Droz-Farny Circles, 69

Euler
  Line, 7, 28
  Points, 6, 46

Fuhrmann Circle, Triangle, 49

Gergonne Point, Triangle, 61, 99, 142

Haruki's Theorem, 144

Isogonal Lines, Points, 53

Lagrange, Joseph, 79
Leibniz, Gottfried, 79
Lemoine Circles, 65, 88, 94

Medial Triangle, 2, 11
Menelaus' Theorem, 147
Miquel's Theorem, 79
  Angle Property, 83

Nagel Point, 5, 7, 12, 50
Nested Triangles, 27
Nine-Point Circle, 6, 117, 127

Orthic Axis, 151
Orthic Triangle, 21, 60
Orthocenter, 17
Orthopole, 125

Parabola, 47
Patruno, 2
Pedal-Cevian Point, 142
Pedal Triangle, Circle, 67, 71, 143

Quadrilateral, Cyclic, 35

Radical Axis, Center, 145
Ramler, O. J., 138

Reflections, 25, 43
Reflector Property, 47
Rigby, John, 30, 44, 50, 59, 63, 65, 73, 141, 166, 170
Rigby Point, 138

Simson Line, 43, 48, 82, 127, 133
Spieker Circle, Center, 3, 7, 12
Splitters, 5
Splitting-center, 5
Steiner Point, 119

Symmedian, Symmedian Point, 53, 106, 122

Tarry Point, 119
Taylor Circle, 95
Trigg, Charles W., 138
Tucker Circles, 87

Wallace, William, 79
Wessel, Caspar, 79
Wilson, John, 79